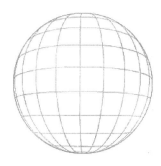

2005 Vol. 61, No. 1

Current Reproductive Technologies: Psychological, Ethical, Cultural and Political Considerations
Issue Editors: Linda J. Beckman and S. Marie Harvey

T0188460

Past JSI Editors

Phyllis Katz (1997–2000)
Daniel Perlman (1993–1996)
Stuart Oskamp (1988–1992)
George Levinger (1984–1987)
Joseph E. McGrath (1979–1983)
Jacqueline D. Goodchilds (1974–1978)
Bertram H. Raven (1970–1973)
Joshua A. Fishman (1966–1969)
Leonard Soloman (1963)
Robert Chin (1960–1965)
John Harding (1956–1959)
M. Brewster Smith (1951–1955)
Harold H. Kelley (1949)
Ronald Lippott (1944–1950)

Journal of Social Issues, Vol. 61, No. 1, 2005, pp. 1–20

Current Reproductive Technologies: Increased Access and Choice?

Linda J. Beckman[*]

Alliant International University, Los Angeles

S. Marie Harvey

University of Oregon, Eugene

This article discusses key issues related to current reproductive technologies including contextual and personal barriers to use, complexity of decision making, limited access to technologies for poor women and women of color, and the politics and social controversy surrounding this area. New reproductive technologies have to be put to the same test as any other product—can and will women use them correctly? We need to not only know about the technology itself; we also need to know about the individuals who intend to use the technology and about contextual factors that influence use. Accordingly, the articles in this issue focus on the multiple determinants that influence acceptability of reproductive technologies and the policy, political, and legal implications associated with their use.

In this issue we define reproductive technologies as the drugs, medical and surgical procedures, and devices that facilitate conception, prevent or terminate pregnancy, and prevent the acquisition and transmission of sexually transmitted infections (STIs). It is important to note that these techniques separate sex from

---

[*]Correspondence concerning this article should be addressed to Linda J. Beckman, Alliant International University, 1000 South Fremont, Unit 5, Alhambra, CA 91803-1360 [e-mail: lbeckman@Alliant.edu].

We dedicate this issue to our mentor and friend, Helen Rodriguez-Trias, M.D., who died on December 27, 2001. Helen's dedication to the fundamental right of women to control their own bodies and her passionate commitment to advancing the health and rights of women, especially poor women and women of color, continues to inspire those of us who were blessed to have had her in our lives. She was an important mentor to women of diverse backgrounds and ages: students, new professionals, established scholars, service providers, and policy advocates. We have learned from her wisdom, that she so willingly shared. We honor her memory through actions that promote the reproductive health of women and improve the quality of their lives.

reproduction (Tangri & Kahn, 1993). Therefore, these technologies allow individuals to engage in sexual intercourse for purposes other than procreation and facilitate procreation without engaging in sexual intercourse.

Our approach to this issue is grounded in our belief that for women to gain equality with men, nationally and internationally, requires that they have control over their bodies and are able to choose whether or not and when to have children. Reproductive health is defined by the World Health Organization (1998) as "complete physical, mental, and social well being in all matters related to the reproductive system" (p. 1). One important strategy for increasing reproductive health is to provide needed services and tools to women to help them overcome infertility; carry wanted pregnancies to term; avoid STIs and prevent unintended pregnancies; when desired, terminate a pregnancy; and enjoy physical and psychological health during and beyond the childbearing years. Our distinctive approach leads to a comprehensive analysis of issues involving reproduction with the goal of promoting more integrated reproductive health services for all women.

Over the last 25 years new reproductive technologies have emerged and extant techniques have been improved or rediscovered. Many new procedures that increase individuals' ability to build a family have led to scenarios previously only visualized in novels such as Huxley's (1998) *Brave New World*. The event that initially galvanized the field of infertility treatment in 1978 was the birth of the first child resulting from in vitro fertilization (IVF), the most popular of the assisted reproductive technologies (ARTs). ARTs are non-coital methods of conception that involve manipulation of both eggs and sperm. The most popular ART is IVF, used in over 70% of all ART procedures (Resolve of Minnesota, n.d.). IVF is a process that uses drugs to stimulate egg production in a woman. The ripened eggs from the ovary are then retrieved, in the laboratory, and fertilized with semen. The resulting embryo or embryos are then transferred back into the uterus for implantation (Centers for Disease Control, 2003; Resolve of Minnesota, n.d.). ARTs are expensive, (averaging $8,000–10,000 per approximately two-week egg retrieval cycle [Resolve of Minnesota, n.d.] and ranging from $60,000 to over $150,000 per successful delivery [Neumann, Gharib, & Weinstein, 1994]). In addition, they are time consuming, involve multiple injections of drugs, and have a modest success rate. Less than 25% of cycles involving fresh, non-donor eggs result in a live birth (American Society for Reproductive Medicine, 2000; Centers for Disease Control, 2003).

Procedures that involve only the use of fertility drugs or intrauterine insemination (IUI), commonly known as artificial insemination (AI), typically are not considered ART. However, for purposes of this issue they are included as reproductive technologies. Both IUI and IVF allow a couple to contract with a third-party woman who carries a child that is genetically linked to one or both members of the couple and who relinquishes that child to the couple after birth. Third party contractual parenting (commonly know as surrogacy) challenges traditional views of what constitutes a family and the relative importance of social versus genetic ties to a child. As discussed in this issue, use of methods such as surrogacy raises profound ethical and legal issues and varies in acceptability by culture. Moreover, because of high costs and lack of insurance coverage, many individuals have limited access to these methods.

In addition to technologies to overcome infertility problems, a host of technological advances are now available to prevent unintended pregnancies and limit unwanted births (Harvey, Sherman, Bird, & Warren, 2002; Schwartz & Gabelnick, 2002; Severy & Newcomer, this issue). Methods to prevent or terminate unwanted pregnancy include female hormones delivered via injection, implant, or pill; mechanical devices placed in the uterus; devices that alert women about their fertile period; and surgical procedures. Moreover, not all methods must be used prior to or during sexual intercourse. Emergency contraception involves the use of hormones up to three to five days after unprotected intercourse to prevent conception. Voluntary termination of pregnancy may involve simple surgical techniques (e.g., electric vacuum aspiration, manual vacuum aspiration) or drug-induced techniques. Some drugs, such as mifepristone (also known as the abortion pill, Mifeprex, or RU 486), have been tested extensively in other countries; others such as methotrexate were originally developed and used for other purposes. Procedures and methods to prevent conception and terminate pregnancies are not nearly as high-tech as those to overcome infertility. Yet, they raise similar types of problems and issues in terms of their acceptability to various cultural and religious groups and because of their limited accessibility. Such problems may be exacerbated by the use of technology for purposes not originally intended (e.g., the use of female hormones originally designed for contraception to control menopausal symptoms and reduce risk of disease in peri-menopausal and postmenopausal women).

Because of the world-wide AIDS pandemic (UNAIDS, 2003) and the high incidence of many other STIs such as chlamydia and gonorrhea nationally and internationally, women and men need methods to protect against Human Immunodeficiency Virus (HIV)/STIs (Eng & Butler, 1997; Rosenberg & Gollub, 1992; Stone, Timyan, & Thomas, 1999).

The male condom is widely recognized as the most effective method of protecting against HIV and some other STIs for sexually active couples (Stone, Timyan, & Thomas, 1999). Some men may, however, be unwilling to use condoms and if women desire protection, they frequently must negotiate condom use with their male partners.

Because of gender-based power inequities, some women may not be able to negotiate condom use to protect themselves against diseases (Amaro, 1995; Amaro & Raj, 2000; Blanc, 2001). There is, therefore, an urgent need for additional, preferably female controlled, methods for HIV/STI prevention. Of critical significance are devices and products still under development such as microbicides (for examples see Koo, Woodsong, Dalberth, Viswanathan, & Simons-Rudolph, this issue; Severy & Newcomer, this issue) that would protect women and their partners from HIV and other STIs. Although these devices and products are designed to prevent disease rather than to control fertility or overcome infertility, issues of acceptability are equally critical to their use.

In this issue we consider psychological, ethical, sociocultural, and political issues of selective technologies. Taken together, the technologies—some old, some new, some still on the horizon—provide more options for women and their partners, theoretically making it possible for them to have greater control over their physical health and psychological well-being. The development of better, more sophisticated scientific technologies generally is viewed by couples and medical professionals as a benefit that could potentially improve physical health and well-being (Kailasam & Jenkins, 2004; Women's Health Weekly, 2004). That said, these technologies have engendered great controversy even among feminists (Henifin, 1993; Tangri & Kahn, 1993) as has their marketing (Kolata, 2002). Feminists have failed to achieve an integrated discourse about women's reproductive decision making across the various technologies (Cannold, 2002). While they support women's right to limit or terminate pregnancy, radical feminists generally oppose assisted reproductive technology. Feminists see women as independent rational decision makers when confronted with an unwanted pregnancy. In contrast, many of them believe that women may be coerced into procedures such an IVF and surrogacy and, therefore, they cannot make unconstrained, independent decisions about these procedures (Cannold, 2002).

Certain religious and cultural groups view some or most of these technologies as unacceptable, even immoral. For instance, the Catholic Church characterizes abortion and contraception as immoral and urges women to forgo these methods (Russo & Denious, this issue; Wakin, 2003). Each of these reproductive technologies raises significant, social, ethical, and psychological issues for women and their sexual partners (e.g., Pasch & Christensen, 2000). Technologies at both ends of the fertility spectrum may be difficult to use and involve significant emotional, social, and/or economic costs (e.g., Pasch & Christensen).

The purpose of this issue is to provide a selective overview of major psychological, ethical, sociocultural, and political issues as they relate to reproductive technologies; to consider the policy implications of these issues; and to promote new research through synthesis and integration of extant literature, the presentation of new data, and identification of new research directions. Prior theoretical development in this area is sparse. Although it is difficult to impose a strong theoretical framework that encompasses the diverse perspectives of these articles, we offer a general conceptual framework to help integrate the multitude of variables examined and issues raised. This framework emphasizes four sets of factors: situational context, relationship context, user characteristics and method characteristics, and their relationships to outcomes associated with reproductive technologies (Table 1). This framework was adapted from an earlier model that focused

Table 1. Conceptual Framework for Reproductive Technologies

Situational Context	Relationship Context	User Characteristics	Method Characteristics	Outcomes
Service delivery system	**Characteristics of the partnership**	**Child-bearing motivation**	**General acceptability**	**Access to Technology**
Accessibility	Stability	**Economic resources**	Difficulty of use	**Was Outcome Achieved?**
Provider attitudes	Stage	SES/Income	Time lost from work or	Proception/Live Birth
Cultural norms/beliefs	Abuse	Education	childcare responsibilities	Contraception/Pregnancy
Procreation	Commitment	Insurance coverage	Side effects	termination
Specific methods	Passion	**Cognitions/Perceptions**	Effectiveness	Disease prevention
Gender roles	Communication	Beliefs about infertility/fertility	Risks	**Psychosocial effects**
Legal system	**Gender roles/Power**	Perceived risk of conception	Sexual enjoyment	Psychological well-being
Political climate	Decision making	Beliefs about specific technology	Other "product"	Relationship with partners
Racism/discrimination	dominance	Perceived stigma	characteristics	**Legal Ramifications**
	Partner support	Sexual comfort	**Female control**	
	Sexual behavior	Attitudes about gender		
		roles/feminism		
		Sociodemographic		
		characteristics		
		Ethnicity		
		Sexual orientation		
		Other		

specifically on factors influencing consistent contraceptive use (Beckman & Harvey, 1996).

Reproductive Technologies: Key Issues

The literature on reproductive technologies, exemplified by articles in this issue, identifies and discusses several key issues and concerns. We discuss these below.

Contextual and Personal Barriers to Use

Many factors restrict women's access to reproductive technology. All four sets of variables identified in Table 1 may serve as barriers to use. For example, lack of uniform statues governing surrogacy (Ciccarelli & Ciccarelli, this issue), lack of health insurance coverage for expensive infertility treatment, lack of information to guide decision making about available options (Woodsong & Severy, this issue), and characteristics of the methods themselves (Harvey & Nichols, this issue; Severy & Newcomer, this issue) may serve as barriers to use. Moreover, barriers are not equitably distributed throughout the social structure (Henifin, 1993). They differentially affect women of certain cultures, race/ethnicities and sexual preferences. Poverty and lack of economic resources, in particular, may limit access to reproductive technology (Henifin, 1993). For example, and as noted above, many technologies, particularly those associated with reversing infertility, are expensive and not covered by medical insurance.

Equally as important as the situational context is the acceptability of a method or procedure to the individual user. New technologies will not be effective in increasing the availability of reproductive health services unless female consumers and providers find these methods acceptable and women are willing to use them. Severy & Newcomer (this issue) describe the issues involved in determining acceptability and the difficulty of measuring it, especially for products not yet in general use. Presumably, the acceptability of pregnancy and disease prevention methods is strongly influenced by the perceived attributes of the specific methods (see Harvey & Nichols, this issue; Koo et al., this issue; Severy & Newcomer, this issue).

In addition, the acceptability of technologies is shaped by factors such as culture, ethnicity, age, social class, and sexual preference. Several articles (e.g., Bird & Bogart, this issue; Harvey & Nichols, this issue; Severy & Newcomer, this issue; Woodsong & Severy, this issue) acknowledge that sociocultural context determines individual perceptions of method or procedure attributes. In other words, method attributes will likely have different meanings and consequences for women, depending on their own personal values and life circumstances.

Complexity of Decision Making

Even when women have ready access to reproductive methods and treatments, the decision to use them can be difficult and emotionally draining. Giving women and their partners more choices also increases the complexity and difficulty of decision making about reproductive issues and may raise painful, ethical, inter-personal, and emotional issues for them (Ciccarelli & Beckman, this issue). In addition, reproductive health decisions frequently are couple decisions rather than individual decisions (e.g., Ciccarelli & Beckman, this issue; Koo et al., this issue; Severy & Newcomer, this issue) which raises issues about power in intimate re-lationships, gender roles, and women's ability to negotiate outcomes with their partners.

Accurate, easily understood information is essential for informed choices and optimal decision making about reproductive options. In the present political cli-mate, accurate information on certain controversial topics such as sexual behavior or abortion may be difficult to obtain. In some cases, curtailment of funding has led to gaps in the knowledge base (Woodsong & Severy, this issue); in others, mis-information may be provided by groups with a specific social agenda or economic interest (Naughton, Jones, & Shumaker, this issue; Russo & Denious, this issue). Even if accurate information is available, decision making may not appear rational to the outside observer. For instance, Naughton and colleagues (this issue) note that despite strong new data about the health-related risks associated with hormone therapy (HT) older women may be reluctant to terminate HT because they believe it helps to promote a youthful appearance.

In addition, culture is of great significance in individual and couple deci-sion making about use of reproductive technology (Burns, 2003; Dugger, 1998; Erickson & Kaplan, 1998). What is acceptable differs depending on cultural values and beliefs (e.g., Harvey, Beckman, & Branch, 2002; Woodsong, Shedlin, & Koo, 2004). Women may desire to postpone or avoid childbearing in order to achieve educational and occupational goals or because they do not have the economic resources to support another child. However, pronatalist norms may propel them toward motherhood (Russo, 1976). Similarly, in part because of these norms and beliefs, women with fertility problems are willing to undergo stressful, painful, expensive, and inconvenient procedures that often are unsuccessful in order to attempt to bear a child of their own (Stanton, Lobel, Sears, & DeLuca, 2002). Still other women and their partners are willing to contract with a stranger in order to have a baby genetically connected to at least one intended parent (Ciccarelli & Beckman, this issue). Most cultural groups identify infertility as a major problem, with especially strong stigma attached to infertility in women (Mabasa, 2002; Remennick, 2000; Whiteford & Gonzalez, 1995). It is ironic that the cultures that are most pronatalist also are ones that most often disapprove of infertility treatments, especially if donated gametes are involved.

The Personal is Political

Reproductive technologies frequently are acclaimed for increasing women's reproductive rights by providing new choices and options (Gollub, 2000; Gollub, French, Latka, Rogers, & Stein, 2001; Latka, 2001; Raphan, Cohen, & Boyer, 2001). Having choice is further extolled as empowering women. To the extent that these technologies allow women (and their partners) more control over the planning and spacing of children, women are more likely to achieve educational and career goals (Zabin & Cardona, 2002) and improve their quality of life and physical health (e.g., Beckman, in press; Murphy, 2003; Wingood & DiClemente, 2000). Moreover, female-controlled pregnancy and STI prevention methods will likely lead to greater feelings of control for the women who use them and provide women with greater ability to avoid unprotected sexual behavior (Stein, 1993, 1995).

We must acknowledge, however, another and darker side to this analysis that (a) suggests that only an elite group of privileged well-to-do non-Hispanic White women can have access to many of these new options, and (b) questions whether even privileged women experience greater control and reproductive autonomy because of the availability of options to promote and prevent conception and birth. It is generally recognized that differential access, especially to costly infertility treatments, exists for privileged versus disadvantaged women (Henifin, 1993). The effects of access to fertility promoting reproductive technology on women's experienced control and empowerment have, however, been relatively ignored. In a recent book McLeod (2002) contends that women's reproductive autonomy in many cases may decline when new reproductive treatments become available. Insensitive treatment by providers, paternalism, and lack of social support may diminish self-trust and thereby threaten a woman's reproductive freedom (McLeod, 2002).

Unfortunately, political agendas and considerations abound as far as reproductive technology is concerned. The politics of reproductive technology often guide social policy, especially in the United States where bitter and sometimes violent struggles have characterized the abortion debate (Beckman & Harvey, 1998; Russo & Denious, this issue; Sherman, this issue). The political controversies, although not as vitriolic in other areas of reproductive technology, extend far beyond abortion and are rooted in differences in religious, moral, and personal values (Russo & Denious, 1998; Russo & Denious, this issue). Characterizing the use of various reproductive technologies as moral issues by conservative religious groups has lead to the promotion of policies that attempt to limit women's access to these technologies. For instance, despite the recommendations provided by two panels of expert advisors, in May 2004 the Food and Drug Administration (FDA) again rejected a pharmaceutical company's application for over-the-counter access to one version of emergency contraceptive pills (ECPs) (Plan B). Although opponents claim that this action was in response to pressure from conservative and anti-abortion groups (American Civil Liberties Union, 2004), spokespersons for the FDA cite lack of research on the safety of ECPs for young adolescents (Food and Drug Administration, 2004).

On the other side of the political spectrum, feminists and advocates for women's rights lobby for social policies to increase access to reproductive technology. In addition, efforts to restrict certain groups' access to technology may be motivated by racist or homophobic beliefs. For instance, ethical questions about lesbians' rights to artificial insemination may be buttressed by homophobic attitudes (Erlichman, 1988; Jacob, 1997). On the other hand, cultural beliefs about government conspiracies can affect the willingness of some African Americans to use specific pregnancy or HIV/STI prevention methods (Bird & Bogart, this issue).

The personal and the political are more deeply interwoven today than in past political landscapes. The executive branch of government in the United States has always used political processes to promote its social agenda. Regrettably, the current Republican administration has attempted to impose its social and moral values upon the American public by removing, withholding, or misrepresenting scientific information when it does not fit their social mandate (Russo & Denious, this issue). Much of this information involves women and girls (National Council of Research on Women, 2004) and nowhere is this attempted embargo on scientific knowledge clearer than in the area of reproductive health issues. Concerned scientists throughout the country have indicated their opposition to both the distortion or withdrawal of scientific information from government reports and Web sites and attempts to subvert the scientific process of merit review in the pursuit of a social agenda (Consortium of Social Science Associations, 2004; Russo & Denious, this issue; Union of Concerned Scientists, 2004). These attempts to limit access to information are detrimental to the mental and physical health of women and to their families and communities (National Council of Research on Women, 2004).

Contents of This Issue

The articles in this issue raise multiple concerns including the best way to deliver services to women (situational context); factors affecting intimate behavior and partner support (relationship context); the influence of user characteristics (e.g., socioeconomic status, culture) as product characteristics of contraceptives and microbicides (e.g., viscosity, color, and smell; method characteristics) on acceptability. The broad range of issues raised stems in part from the diversity of the technologies examined.

With the exception of this overview, we have divided the contributions into two distinct groups: (a) Acceptability and the Importance of Individual and Contextual

Factors; and (b) Policy, Political, and Legal Implications. These articles approach the issues from varying perspectives; two (Harvey & Nichols, this issue; Koo et al., this issue) present original data, several (e.g., Russo & Denious, this issue; Sherman, this issue; Woodsong & Severy, this issue) focus primarily on politics and policy, and four (Ciccarelli & Beckman, this issue; Ciccarelli & Ciccarelli, this issue; Russo & Denious, this issue; Sherman, this issue) devote significant attention to legal and ethical issues. Although several of the articles are integrative reviews, the authors comment briefly on policy implications and/or ethical and legal dilemmas.

The articles in Section II consider multiple determinants (see Table 1) associated with reproductive technologies designed to prevent or terminate unintended pregnancy, allow infertile couples to build families, and protect against the transmission of STIs including HIV. Particular emphasis is placed on the acceptability of these technologies to current and potential users.

ARTs help infertile couples to successfully conceive and deliver children or to contract with a third party to achieve a birth (American Society for Reproductive Medicine, 2000; Resolve of Minnesota, n.d.). At the same time, their use raises difficult issues such as what determines a family and the possible exploitation of poorer women for the benefit of wealthier individuals, as well as a quagmire of ethical and legal concerns. The legal, ethical, and social dilemmas raised by the use of ARTs may frequently give rise to exceedingly difficult personal decisions and high stress levels among participants (Pasch & Christensen, 2000). Moreover, the relationship context (e.g., social support, relative power) as well as the characteristics of ARTs often make their use demanding and stressful (Abbey, 2000; Pasch & Christensen, 2000). From among the many types of assisted reproductive technologies, we have chosen to highlight the most complex and controversial of these methods, third party assisted reproductive technology commonly known as surrogacy.

The first article, by Ciccarelli and Beckman (this issue), discusses various aspects of surrogacy arrangements. The authors summarize the meager social and psychological research currently available. The relevant research literature primarily focuses on attitudes toward surrogacy and the motivations, experiences, and traits of surrogate mothers and their perceptions of the surrogacy process. Studies of intended parents and the children resulting from surrogacy arrangements are, for the most part, notably absent. Although surrogacy is perceived as the least acceptable of the ARTs, the limited empirical literature generally does not support concerns about the emotional damage or exploitative nature of surrogacy for surrogate mothers or intended parents. The social and psychological complexities of having a child increase dramatically when more than two people are necessary to produce a child and there are many points at which relationships among participants can be difficult. Although surrogacy arrangements are quite rare, they appear to be here to stay. Therefore, additional research could provide useful information that would serve as a foundation for evidence-based counseling during all phases of the surrogacy process.

Pregnancy prevention and STI prevention are inextricably intertwined because the same behaviors put individuals at risk of both outcomes. In addition, as previously mentioned the male condom, currently the most effective method for prevention of HIV/STIs among sexually active individuals, is also an effective contraceptive method (Stone, Timyan, & Thomas, 1998). The next two articles consider the acceptability of new technologies for prevention of both outcomes. Severy and Newcomer (this issue) discuss the determinants of product acceptability that apply both to contraceptives and methods that prevent the transmission of HIV and other STIs. For these authors the key question is how best to inform the development and dissemination of effective methods for pregnancy or disease protection that are acceptable to those whose sexual behavior places them at risk. They discuss, provide examples of, and critique the three models that past research on acceptability has typically followed: hypothetical product acceptability, embedded behavioral studies within clinical trials, and post-marketing studies. Post-marketing research has been particularly effective in studying the interplay among the individual user, product characteristics and the larger social context (i.e., the health care system) in determining acceptability. These studies examine individual product characteristics and interpersonal factors that may influence acceptability.

In particular, the authors acknowledge that decision making involves two partners—acting in a dynamic fashion and influencing each other's acceptability and behavior. The authors contend that considerable evidence indicates that acceptability is a function of partners' beliefs about the impact of the method on their intimate relationship and sexual enjoyment. Accordingly, Severy and Newcomer review previous research that examines the influence of contraceptive methods on sexual behavior and enjoyment as major determinants of acceptance. They conclude with a review and critique of seven "new" technologies (e.g., vaginal rings, microbicides, and devices that inform couples of the woman's fertile period) with a focus on acceptability issues discussed earlier in the article.

Koo, Woodsong, Dalberth, Viswanathan, & Simons-Rudolph (this issue) provide important information about the dyadic nature of reproductive decision making. More specifically, they examine the importance of the social context for acceptability of topical microbicide use within sexual relationships. As Koo et al. point out, the nature of the sexual relationship influences perceptions of risk and, thus, need for protection, as well as the willingness and ability to use a microbicide. Data from multiethnic samples of women and their male partners demonstrate that perceptions of risk differ by relationship type. Partners in new relationships are more likely to acknowledge the need for protection because they have no expectation of sexual exclusivity. The norms espoused by both female and male participants that partners in serious relationships should talk about STI/HIV protection present a paradox for women. Men also believed that women in serious relationships should not need to use protection; therefore, talking about protection was perceived as problematic and potentially damaging because it could raise partner suspicion and decrease relationship trust. Therefore, it is not surprising that both men and women reported that communication about STI/HIV prevention usually did not occur in steady relationships.

Much of the interest in topical microbicides has been spurred in part by the possibility of their covert use by women. Rather than attempting to hide method use, some women suggested that it would be desirable to hide the purpose for which the product was used by telling their partner that the microbicide provided protection against pregnancy or could enhance sexual pleasure (as well as provide HIV/STI protection). In contrast, men generally opposed direct or indirect covert product use by women, in part because they were concerned about product characteristics and side effects. The Koo et al. article illustrates the complexity of the determinants of method acceptability when dyadic processes are involved and the potential method is still under development.

The development of an instrument to assist health care providers in determining which characteristics of abortion methods are important to a woman would be helpful in matching women with a method that they find acceptable. The Harvey and Nichols (this issue) article describes the development and evaluation of the Abortion Attributes Questionnaire (AAQ), a self-administered research instrument designed to assess the perceived importance of specific characteristics of abortion methods (i.e., Method Characteristics in Table 1). The attributes included in the AAQ were derived from a review of the literature that examined reasons women report for choosing an abortion method and women's assessment of perceived advantages and disadvantages of medical and surgical abortion.

The authors administered the questionnaire to sample of women who chose either a medical or surgical abortion procedure for pregnancy termination. They were asked to rate how important each of 21 characteristics would be "when choosing between surgical and medical abortion." A factor analysis yielded four distinct factors, which reflect salient characteristics of abortion methods: absence of side effects, natural/avoidance of surgery, effectiveness and timing, and personal control. As subscales the factors were found to have good internal reliability. In addition, results indicated that three of the four factors predicted choice of abortion method. The authors conclude with a discussion of the implications of the AAQ as an instrument for both clinical applications and applied research.

The rates of HIV, other STIs and unintended pregnancy are disproportionably high in the African American population. Bird and Bogart (this issue) review the literature on conspiracy beliefs regarding HIV/AIDS and birth control among African Americans and discuss the policy and programmatic implications of such beliefs for the prevention of HIV, other STIs, and unintended pregnancy. Research has documented the prevalence of conspiracy beliefs among African Americans, but almost no previous work has examined the association between conspiracy beliefs and risky sexual behavior or acceptability of HIV treatment. They specifically address the impact of conspiracy beliefs on sexual attitudes and behaviors relevant to prevention of HIV, other STIs, and unintended pregnancy.

Findings from the authors' small telephone survey of African Americans (Bird & Bogart, 2003; Bogart & Bird, 2003) suggest that endorsement of conspiracy beliefs is associated with attitudes towards condoms and other contraceptive methods, contraceptive intention and choice, and sexual behavior. It is noteworthy that different patterns of results were found for two types of HIV/AIDS conspiracy beliefs: beliefs about the government's role in the AIDS epidemic and treatment-related conspiracy beliefs. Conspiracy beliefs can affect the acceptability of disease and pregnancy prevention methods currently available. In addition, such beliefs may pose significant barriers to the use of new methods and antiretroviral treatment regimens for HIV+ individuals. Bird and Bogart's article (this issue) illustrates how the acceptability of reproductive technologies among particular populations can be affected by cultural beliefs shaped by a specific historical and social context.

The articles in Section III describe the legal and public policy implications of various reproductive technologies. Many of these reproductive health procedures and products challenge traditional views about the nature of sexuality and appropriate sexual behavior. As described in the articles, the conflict between traditional values and reproductive technology has led to legal ambiguities and political manipulations.

Ciccarelli and Ciccarelli (this issue) is a companion piece to the Ciccarelli and Beckman (this issue) integrative review of the psychological aspects of surrogacy arrangements. Ciccarelli and Ciccarelli highlight the myriad of legal issues and inconsistent statutes and legislation involving surrogacy. Legal controversies underscore the need for uniform laws that lessen uncertainty for intended parents and surrogate mothers, thereby alleviating their fear and anxiety about surrogacy agreements. Ciccarelli and Ciccarelli point out that the vast majority of states have no laws on the issue of surrogacy, with state laws that do exist running the entire spectrum from surrogacy violating criminal statutes to surrogacy-supportive. The major distinction is whether the surrogacy involves IVF with both of the intended parents' gametes or IUI in which the intended father's sperm is used for insemination. In the case of IVF, most states that have laws recognize the intended parents as the legal parents upon conception, and either implicitly or explicitly recognize the validity of a contract. In IUI, adoption law generally applies. In practical effect, any contracts or agreements that are signed prior to birth will be completely unenforceable and the surrogate will have some time period to change her mind about whether she will relinquish custody of the child.

The next four articles examine from differing perspectives the impact of the values, beliefs, and social agendas of stakeholders on public policy involving reproductive technologies. Sherman (this issue) reviews the literature and events related to the politicization of emergency contraceptive pills (ECPs) and their availability in the United States. She contends that the basis for opposition to emergency contraception is rooted in the modes of action of ECPs and the confusion of this regimen with medical abortion. Because of its multiple modes of action, (i.e., delaying ovulation; preventing fertilization; and preventing implantation of a fertilized egg) some anti-abortion groups have label ECPs as an abortifacient and have mobilized their opposition. Emergency contraception involves use of hormonal contraceptives within 72 to 120 hours of unprotected intercourse, whereas drugs that induce an abortion are administered after confirmation of pregnancy but early in the gestational period. While medical abortifacients cause an implanted egg to be expelled from the uterus, ECPs are ineffective once implantation has occurred.

Sherman reviews legislative actions that could potentially limit the availability of ECPs. In addition, she discusses current efforts to increase access and availability through innovative programs, legislation, and changes in medical practice. After providing evidence of the acceptability of ECPs to both consumers and health care providers, Sherman concludes with a call for action. This call includes recommendations for future research, the delivery of services, and public policy.

Naughton, Jones, and Shumaker (this issue) discuss another controversial application for female hormonal agents, hormone therapy (HT) for peri-menopausal and postmenopausal women. They examine the history of hormonal therapy and menopause from the point of view of competing stakeholders in the process, including women users, health providers, the pharmaceutical industry, and the medical research establishment. The different and often competing objectives of these groups have led to health policies and practices that do not benefit women and in some cases may harm them. Although HT originally was designed as a short-term therapy to alleviate unpleasant symptoms associated with the menopause, beginning in the 1980s it became popular for the prevention and treatment of major diseases prevalent among older women, namely heart disease, osteoporosis, and dementia. However, in the 1990s research began to suggest that HT involved possible risk, particularly increased risk of breast cancer. In 2002 the estrogen plus progestin HT clinical trial of the Women's Health Initiative was terminated early because of data suggesting increased risk of breast cancer, dementia, and other diseases. Moreover, the trial showed no benefits of HT in prevention of heart disease and some increase in risk during the first year of use. More recently in 2004 the estrogen-only arm of the trial was similarly terminated.

These results have produced great controversy; Naughton et al. contend that this controversy exists because some of the stakeholders, such as major pharmaceutical manufacturers, find the results difficult to accept. The drug companies have already incurred financial loss, are concerned about further loss of profits, and are under pressure to produce for their shareholders. Physicians experience dissonance because scientific results are in conflict with their practice patterns and beliefs about drug effects. They also often are overwhelmed by the large amount of information available, have limiting training in research methods, and may rely on drug company representatives as a primary information source on treatment agents. Aging women users who have incorporated the belief that HT can combat the effects of aging on physical appearance may be reluctant to forgo the treatment regardless of increased risk.

The article by Russo and Denious (this issue) highlights the political context surrounding some new and extant reproductive technology. The authors contend that research results on abortion and contraception are frequently misrepresented. They attribute this misrepresentation to a right wing conservative social agenda that seeks to restrict availability and access to reproductive technology, thereby limiting women's options. According to the authors, the contention that abortion is associated with major negative physical and mental health effects, despite lack of supportive scientific evidence, is harmful to women's health and psychological well-being. The moralistic approach of conservative groups is reflected, also, in legislation and executive orders regarding the global AIDS problem, in HIV prevention and abortion funding, and in the manipulation of the federal scientific review advisory process. Russo and Denious question what responsible scientists can and should do about the current sociopolitical context. They conclude that scientists need to strongly defend against political interference in the science advisory process. In addition, they need to monitor the dissemination of the results of their research beyond what is reported in scientific journals to ensure that findings are not misrepresented. Dissemination of accurate research findings requires immediate responses to misrepresentations of scientific data in the popular media, the development of new outlets for scientific knowledge such as Internet Web sites, and working with professional organizations and groups to educate the public.

The final article (Woodsong & Severy, this issue) further discusses the importance of provision of accurate scientific information to the public. Women and their partners as well as health providers need accurate and accessible information about reproductive technology options to use in their decision making process. Yet, there is a paucity of scientific information available on reproductive and sexual health topics. Using the example of topical microbicides, the authors identify gaps in providers' knowledge that may limit acceptability of this emerging technology and fail to inform reproductive decision making of patients. Lack of knowledge, they contend, can influence the supply, demand, and correct use of reproductive health options. Accurate data help women to understand the risks and benefits of different reproductive technology options. These data need to reflect effectiveness of use in real life circumstances and different sexual situations and practices rather than the efficacy of clinical trials.

Woodsong and Severy's use the concept of "effectiveness equipoise," a case in which one treatment or intervention does not appear to have a clear advantage over the others and "partial effectiveness" to describe the complexity of the microbicide decision making. Also, they emphasize the important of individual and cultural values in choices among reproductive health options. In recent years research has been limited because of funding constrains in both the public sector and the private sectors. An even more alarming trend is that some topics such as condom efficacy and health risks of abortion have been pegged as controversial and information previously available on such topics has been misrepresented or removed from government Web sites.

Conclusion

The contents of this collection of articles is determined both by areas of particular concern to the editors and by the expertise of the authors. By making choices about what to include and how to conceptualize these articles, we made implicit decisions to de-emphasize many equally important content areas and approaches. We would like to acknowledge a few of these potential limitations. First, although the role of the male partner is considered, we consciously chose to primarily emphasize access to and the effects of these technologies for women because women are the ones who carry and bear children. Thus, we do not include literature on male infertility or the effects of reproductive technologies for males. Second, while we recognize the importance of technology to overcome infertility, the vast majority of contributions to this issue discuss pregnancy control and protection against disease. Third, even though we emphasize the effects of cultural difference, we do not explicitly take a developmental approach; instead we concentrate largely on adults. Thus, a vast literature on adolescent pregnancy and HIV/STI prevention is not considered. Fourth, we consciously chose to avoid controversial methods of reproduction such as cloning that, as of this time, have not lead to the live birth of humans (The Lancet, 2003; Raeburn, Keenan, & Weintraub, 2002). Fifth, because of our limited expertise in the area of prenatal care, we have focused on technologies that alleviate the inability to conceive rather than those that prevent miscarriage or spontaneous abortion among women who have difficulty carrying a pregnancy to term. Finally, the articles largely focus on research in the United States and national policy implications. An international perspective is not provided and we assume that cultural, social, and economic barriers to women's access to reproductive technology are significantly different in other parts of the world.

In summary, as new reproductive technologies are developed and become available to the public, scientific debate and popular media coverage are often obsessed with the scientific advances these new technologies represent. But what do these technologies mean to everyday women and men who are meant to be the beneficiaries of these advances? As the following articles demonstrate, the issues related to the use of new and extant reproductive technologies often have nothing to do with technology itself. All new reproductive technologies, even the most impressive, have to be put to the same tests as any other product—can and will people use them successfully? In order to answer this question, we need to know something about the technology itself; but, more importantly, we need to know more about the individuals who intend to use the technology. We also need to better understand the situational constraints on access to the technology and to develop better tools to counter campaigns that in the service of a particular social agenda are harmful to women's reproductive health. The articles that comprise this issue provide a fuller understanding of the multiple factors that influence individuals to use or not use certain reproductive technologies and why.

References

Abbey, A. (2000). Adjusting to infertility. In J. D. Harvey, & E. D. Miller (Eds.), *Loss and trauma: General and close relationship perspectives* (pp. 331–344). Ann Arbor, MI: Edwards Brothers.

Advances in fertility technology open new doors for couples. (2004, February 19). *Women's Health Weekly*, pp. 98–99. Retrieved June, 22, 2004, from Health Source—Consumer Edition database.

Amaro, H. (1995). Love, sex, and power: Considering women's realities in HIV prevention. *American Psychologist, 50*, 437–443.

Amaro, H., & Raj, A. (2000). On the margin: Power and women's HIV risk reduction strategies. *Sex Roles, 42*, 723–749.

American Civil Liberties Union. (2004, May 12). *Food and Drug Administration puts politics before women's health; ACLU says investigation warranted*. Retrieved June 22, 2004, from http://www.aclu.org/ReproductiveRights/ReproductiveRights.cfm?ID=15690&c=225

American Society for Reproductive Medicine. (2000). *Fact Sheet: In Vitro Fertilization (IVF)*. Retrieved May 20, 2003, from http://www.asrm.org/Patients/FactSheets/invitro.html

Beckman, L. J. (in press). Women's reproductive health: Issues, findings and controversies. In C. Goodheart & J. Worell (Eds.), *Handbook of girls' and women's psychological health*. New York: Oxford University Press.

Beckman, L. J., & Harvey, S. M. (1996). Factors affecting the consistent use of barrier methods of contraception. *Obstetrics and Gynecology, 88*(Suppl.), 65–71.

Beckman, L. J., & Harvey, S. M. (Eds.). (1998). *The new civil war: The psychology, culture and politics of abortion*. Washington, DC: American Psychological Association.

Bird, S. T., & Bogart, L. M. (2003). Birth control conspiracy beliefs, perceived discrimination and contraception among African Americans: An exploratory study. *Journal of Health Psychology, 8*, 263–276.

Blanc, A. K. (2001). The effect of power in sexual relationships on sexual and reproductive health: An examination of the evidence. *Studies in Family Planning, 32*, 189–213.

Bogart, L. M., & Bird, S. T. (2003). *Relationship of conspiracy beliefs about HIV/AIDS to sexual behaviors and attitudes among African American adults*. Manuscript submitted for publication.

Burns, L. H. (2003). *Cross-cultural issues and the use of assisted reproductive technologies*. Unpublished manuscript.

Cannold, L. (2002). Understanding and responding to anti-choice women-centred strategies. *Reproductive Health Matters, 10*(19), 171–179.

Centers for Disease Control. (2003). 2000 *Assisted Reproductive Technology success rates: Commonly asked questions about U.S. ART clinic reporting system*. Retrieved May 20, 2003, from the National Center for Chronic Disease and Prevention and Health Promotion Web site: http://www.cdc.gov/nccdphp/drh/ART00/faq.htm

Consortium of Social Science Associations. (2004). *Coalition to protect research: Support scientific integrity*. Retrieved June 22, 2004, from http://www.cossa.org/CPR/scientificintegrity.html

Dugger, K. (1998). Black women and the question of abortion. In L. J. Beckman & S. M. Harvey, (Eds.), *The new civil war: The psychology, culture and politics of abortion*. (pp. 107–131). Washington, DC: American Psychological Association.

Eng, T., & Butler, W. T. (1997). *The hidden epidemic: Confronting sexually transmitted diseases.* Washington, DC: National Academy Press.

Erickson, P. L., & Kaplan, C. P. (1998). Latinas and abortion. In L. J. Beckman & S. M. Harvey (Eds.), *The new civil war: The psychology, culture and politics of abortion.* (pp. 133–155). Washington, DC: American Psychological Association.

Erlichman, K. L. (1988). Lesbian mothers: Ethical issues in social work practice. *Women and Therapy, 8,* 207–224.

Food and Drug Administration. (2004, May 7). *FDA issues not approvable letter to Barr Labs; Outlines pathway for future approval.* Retrieved June 22, 2004 from, http://www.fda.gov/bbs/topics/news/2004/NEW01064.html

Gollub, E. L. (2000). The female condom: Tool for women's empowerment. *American Journal of Public Health, 90,* 1377–1381.

Gollub, E. L., French, P., Latka, M., Rogers, C., & Stein, Z. (2001). Achieving safer sex with choice: Studying a women's sexual risk reduction hierarchy in an STD clinic. *Journal of Women's Health & Gender-Based Medicine, 10,* 771–783.

Goodbye dolly . . . and friends? (2003) *The Lancet, 361*(9359), 711.

Harvey, S. M., Beckman, L. J., & Branch, M. R. (2002). The relationship of contextual factors to women's perceptions of medical abortion. *Health Care for Women International, 23,* 654–665.

Harvey, S. M., Sherman, S. A., Bird, S. T., & Warren, J. (2002). *Understanding medical abortion: Policy, politics and women's health* (Policy Matters Paper 3). Eugene: University of Oregon, Center for the Study of Women in Society.

Henifin, M. S. (1993). New reproductive technology: Equity and access to reproductive health care. *Journal of Social Issues, 49,* 61–74.

Huxley, A. (1998). *Brave New World* (Perennial Classics ed.). New York: Harper Collins. (Original work published in 1932).

Jacob, M. C. (1997). Concerns of single women and lesbian couples considering conception through assisted reproduction. In. S. R. Leiblum (Ed.), *Infertility: Psychological issues and counseling strategies.* (pp. 189–206). Oxford, England: Wiley.

Kailasam, C., & Jenkins, J. (2004). Risks and benefits of assisted conception. *Pulse, 64*(5), 46–47. Retrieved June 22, 2004, from the Health Source: Nursing/Academic Edition database.

Kolata, G. (2002, January 1). Fertility Inc: Clinics race to lure clients. *New York Times,* pp. D1, D7.

Latka, M. (2001). Female-initiated barrier methods for the prevention of STI/HIV: Where are we now? Where should we go? *Journal of Urban Health, 78,* 571–580.

Mabasa, L. F. (2002). Sociocultural aspects of infertility in a Black South African community. *Journal of Psychology in Africa, South of the Sahara, the Caribbean and Afro-Latin America, 12,* 65–79.

McLeod, C. (2002). *Self-trust and reproductive autonomy.* Cambridge, MA: MIT Press.

Murphy, E. M. (2003). Being born female is dangerous for your health. *American Psychologist, 58,* 205–210.

National Council for Research on Women. (2004). *Missing: Information about women's lives.* Retrieved May 12, 2004, from http://www.ncrw.org/misinfo/report.pdf

Neumann, P. J., Gharib, S. D., & Weinstein, M. C. (1994). The cost of a successful delivery with in vitro fertilization. *New England Journal of Medicine, 331,* 239–243.

Pasch, L. A., & Christensen, A. (2000). Couples facing fertility problems. In K. B. Schmaling & T.G. Sher (Eds.), *The psychology of couples and illness: Theory, research and practice* (pp. 241–267). Washington, DC: American Psychological Association.

Raeburn, P., Keenan, F., & Weintraub, A. (2002). Everything you need to know about cloning [Electronic version]. *Business Week, 3780,* 44–45. Retrieved June 22, 2004, from the Academic Search Premier database.

Raphan, G., Cohen, S., & Boyer, A. M. (2001). The female condom, a tool for empowering sexually active urban adolescent women. *Journal of Urban Health, 78,* 605–613.

Remennick, L. (2000). Childless in the land of imperative motherhood: Stigma and coping among infertile Israeli women. *Sex Roles, 43,* 821–841.

Resolve of Minnesota. (n.d.). *Assisted reproductive technologies.* Retrieved May 20, 2003, from http://www.resolvemn.org/art.htm

Rosenberg, M. J., & Gollub, E. L. (1992). Commentary: Methods women can use that may prevent sexually transmitted disease, including HIV. *American Journal of Public Health, 82*, 1473–1478.

Russo, N. F. (1976). The motherhood mandate. *Journal of Social Issues, 32*, 143–154.

Russo, N., & Denious, J. (1998). Why is abortion such a controversial issue in the United States. In L. J. Beckman & S. M. Harvey (Eds.), *The new civil war: The psychology, culture and politics of abortion* (pp. 25–59). Washington, DC: American Psychological Association.

Schwartz, J. L., & Gabelnick, H. L. (2002). Current contraceptive research. *Perspectives on Sexual and Reproductive Health, 34*, 310–316.

Stanton, A. L., Lobel, M., Sears, S., & DeLuca, R. S. (2002). Psychosocial aspects of selected issues in women's reproductive health: Current status and future directions. *Journal of Consulting and Clinical Psychology, 70*, 751–770.

Stein, Z. (1993). HIV prevention: An update on the status of methods women can use. *American Journal of Public Health, 83*, 1379–1382.

Stein, Z. (1995). Editorial: More on women and the prevention of HIV infection. *American Journal of Public Health, 85*, 1485–1487.

Stone, K. M., Timyan, J., & Thomas, E. L. (1998). Barrier methods for the prevention of sexually transmitted diseases. In K. K. Holmes, P. Mardh, P. F. Sparling, S. M. Lemon, W. E. Stamm, & P. Piot (Eds.), *Sexually transmitted diseases*. New York: McGraw-Hill.

Tangri, S. S., & Kahn, J. R. (1993). Ethical issues in the new reproductive technologies: Perspectives from feminism and the psychology profession. *Professional Psychology: Research and Practice, 24*, 271–280.

UNAIDS. (2003). *Report on the global HIV/AIDS epidemic: 2002*. Geneva, Switzerland: Author.

Union of Concerned Scientists. (2004, March). *Scientific integrity in policymaking: An investigation into the Bush Administration's misuse of science*. Retrieved June 22, 2004, from http://www.ucsusa.org/publications/report.cfm?publicationid=730

Watkin, D. J. (2003, November 13). Bishops open a new drive opposing contraception. *New York Times*, pp. A20.

Whiteford, L. M., & Gonzalez, L. (1995). Stigma: The hidden burden of infertility. *Social Science and Medicine, 40*, 27–36.

Wingood, G. M., & DiClemente, R. J. (2000). Application of the Theory of Gender and Power to examine HIV-related exposures, risk factors, and effective interventions for women. *Health Education & Behavior, 27*, 539–565.

Woodsong, C., Shedlin, M., & Koo, H. (2004). Natural, normal and sacred: Beliefs influencing the acceptability of pregnancy and STI/HIV prevention methods. *Journal of Culture, Health and Sexuality, 6*, 67–88.

World Health Organization. (1998). *Division of reproductive health: Overall aims and goals*. Geneva, Switzerland: Author.

Zabin, L. S., & Cardona, K. M. (2002). Adolescent pregnancy. In G. Wingood & R. DiClemente (Eds.), *Handbook of women's sexual and reproductive health* (pp. 231–253). New York: Kluwer Academic/Plenum.

LINDA J. BECKMAN is a Professor in the Clinical Psychology PhD program of the California School of Professional Psychology at Alliant International University. She is an associate editor of the *Psychology of Women Quarterly* and author of numerous articles on women's reproductive health.

S. MARIE HARVEY is the Director of the Research Program on Women's Health at the Center for the Study of Women in Society and an Associate Professor of Public Health at the University of Oregon. She is a Fellow of the American Psychological Association (APA) and Past-President the APA Division 34, Population and Environmental Psychology and Past-Chair of the Population, Family Planning and Reproductive Health Section of the American Public Health Association. Her current research interests focus on the acceptability of reproductive technologies; the prevention of HIV/STIs among high-risk women, men, and couples; and the influence of relationship factors on sexual risk-taking. Marie currently serves as Principal Investigator on "Women's Acceptability of the Vaginal Diaphragm," a three year study recently funded by the National Institute for Child Health and Human Development.

Journal of Social Issues, Vol. 61, No. 1, 2005, pp. 21–43

Navigating Rough Waters: An Overview of Psychological Aspects of Surrogacy

Janice C. Ciccarelli*

Claremont, California

Linda J. Beckman

Alliant International University, Los Angeles

This article provides an overview of the social and psychological aspects surrounding the surrogacy process including attitudes about surrogacy, perceptions and problems of surrogate mothers and intended/social parents, and questions concerning children resulting from contractual parenting. Review of the literature on contractual parenting reveals a wealth of discussion about the ethical, moral, legal, and psychological implications, but limited empirical data on the psychological and social aspects. Future research can provide empirical evidence as a foundation for counseling at all phases of the surrogacy process.

Surrogacy is both the oldest and the most controversial of reproductive innovations. Its documented history goes back at least as far as the Old Testament in which Hagar begot Ishmael with Abraham after his wife, Sarah, failed to conceive (Gen., 16 Authorized [King James] Version). Moreover, artificial insemination (AI), a widely used method for surrogacy arrangements, is neither new nor high tech. It has been available for more than 100 years (Hammer-Burns & Covington, 1999, p. 20) and can be performed without medical assistance using a simple turkey baster (Ciccarelli, 1997; Gallagher, 1989). In the last 25 years, however the commercialization of surrogate mothering and the media firestorm associated with the Baby M case (Matter of Baby M, 1988) have led to a groundswell of interest and controversy about this technology (Ciccarelli, 1997).

---

*Correspondence concerning this article should be addressed to Janice C. Ciccarelli, 250 W. 1st Street, Suite 314, Claremont, California 91711 [e-mail: DOCCHICK@msn.com].

Contractual parenting (commonly know as surrogacy) occurs when a couple, the intended parents, contracts with a woman to carry a child for them and to relinquish that child to them after birth (Ciccarelli, 1997; Ragone, 1996). There are two major types of surrogacy arrangements: traditional surrogacy and gestational surrogacy. In traditional surrogacy, the surrogate is impregnated with the sperm of the male partner of the intended parents. In this case, the impregnated woman is both the genetic and birth (i.e., gestational) mother and the intended father is also the genetic father (Ciccarelli, 1997; Ragone, 1996). Gestational carrier surrogacy is used when the female partner of the intended couple has viable eggs but is unable to successfully carry a pregnancy to term. The intended mother's eggs are fertilized with her male partner's sperm in the laboratory using in vitro fertilization (IVF) and the embryo is then implanted in the "surrogate" mother's uterus. In gestational surrogacy, the woman who carries the child has no genetic connection to the child and the intended parents are also the genetic parents (Ciccarelli, 1997; Ragone, 1996).

Some feminist writers have objected to the social construction of the woman who carries the child as the *surrogate* or *surrogate mother*. They contend that such terms do not accurately reflect the reality of contractual parenting since the pregnant woman is the *actual* mother, that is, the gestational or birth mother. Current terminology, they believe, minimizes the value of the gestational mother's role (Hanafin, 1999; Tangri & Kahn, 1993) and delegitimizes her right to a continuing relationship with the child (Jaggar, 1994, p. 379). These issues are important to acknowledge. However, surrogate motherhood reflects the intent of the gestational mother and how she perceives herself and her role (Hanafin, 1999). This term also allows us to distinguish women who bear a child as a result of contractual parenting from other birth mothers. Therefore, we will use the term traditional surrogate for the woman who conceives via AI using the sperm of the father who intends to rear the child and the term gestational surrogate for the woman who carries an embryo that has been conceived via IVF using the intended parents' egg and sperm. The couple that contracts with the surrogate mother is referred to as the intended, social, commissioning or contracting parents, depending on where they are in the surrogate parenting process.

As one can well imagine, the social, psychological, and legal complications increase dramatically as the number of people necessary to conceive a child is increased from the traditional two people (Ciccarelli, 1997).

Review of the literature on contractual parenting reveals a wealth of discussion about the ethical, moral, legal, and psychological implications, but limited empirical data on the psychological and social aspects. Discussion of surrogacy has been ripe with controversy and has assembled some unusual allies. Religious fundamentalists, the Roman Catholic church, and feminists alike have condemned the practice of contractual surrogacy as "baby selling"—one that demeans and threatens women.(e.g., Gibson, 1994; Macklin, 1988; Rothman, 1989; Raymond,

1998; Tangri & Kahn, 1993). The level of controversy engendered by surrogacy, is reminiscent of the abortion controversy in the United States. Surrogacy, like abortion, is controversial precisely because it evokes and often contradicts basic concepts about family, motherhood, and gender roles (Luker, 1984). Conservative groups are fearful that surrogacy will undermine traditional cultural values about the two-parent family with wife primarily responsible for childcare and husband as provider and patriarch (Burr, 2002). On the other hand, many feminists are alarmed about the commodification of women (Tangri & Kahn, 1993) and both groups deplore contractual surrogacy as the selling of babies. Few issues have so deeply divided the feminist community (Behuniak-Long, 1990; Taub, 1992). Pitted against the large group of feminists who oppose contractual surrogacy are others who fear that any limitation of women's reproductive freedom will provide inroads toward curtailment of women's reproductive rights by groups, often religious in nature, that are opposed to women's access to abortion and contraception (e.g., Bartholet, Draper, Resnik, & Geller, 1994; Mahoney, 1988).

Given the level of controversy engendered, one might expect considerable research activity. Yet the research literature is extremely sparse for a number of reasons. First, the absence of funded research on the topic suggests that financial support for research on such a controversial issue may be difficult to secure. Governmental support may be absent when a practice (e.g., abortion, surrogacy) conflicts with the policy of the administration in power. Second, despite the flood of media attention, particularly in the late 1980s and early 1990s, surrogacy arrangements are less common than generally perceived. Historically, there has been no way to track the number of children born as a result of AI. However, since 1992 federal law has mandated that fertility clinics track and report statistics relating to IVF cycles and births (Fertility Clinic Success Rate and Certification Act). The first compilation of these statistics was published by the Centers for Disease Control (CDC, n.d.) in 1995. Unfortunately, this mandate did not include segregating the number of IVF surrogacy births from the total of IVF births. Reporting on IVF surrogacy births became a requirement for fertility clinics in 2003.

Nonetheless, the American Society for Reproductive Medicine has attempted to compile information regarding IVF surrogacy and non-surrogacy births prior to the enactment of the law. According to their statistics, from 1985 through 1999 there were 129,000 babies born as a result of IVF. From 1991 through 1999 there were 1600 babies, included in this total, who were born as result of IVF surrogacy (American Society of Reproductive Medicine, personal communication, June, 2002). The numbers pertaining to IVF births, including surrogacy births, may be low since, prior to enactment of the above mentioned act in 1992, reporting was voluntary. Further, until 2003 reporting regarding surrogacy still was voluntary. In any event, it is clear that contractual parenting is infrequent in comparison with the overall birth rate, even for birth rates involving assisted reproductive technologies.

Third, given the social stigma associated with surrogacy, parties to surrogacy agreements, particularly the contracting couple, relish their privacy and therefore may be unlikely to agree to participate in research (Ciccarelli, 1997; Ragone, 1996). In addition, those who arrange contracts and counsel the parties involved are committed to protecting their privacy for ethical and legal reasons. Low prevalence of surrogacy arrangements and concerns about privacy have led to limited availability of research participants, especially intended parents.

Research information is important to clinical psychologists and other mental health providers because it is difficult to screen, advise, and counsel both surrogate mothers and intended parents if there are no empirical bases for such professional activities, (Hanafin, 1999). Due to lack of empirical data on surrogacy screening and counseling, some clinicians have attempted to glean data from the adoption literature for use in surrogacy. Such comparisons appear inadequate since surrogacy is exceedingly more complex than adoption and has many fewer government laws and regulations structuring it (Hughes, 1990). Research about the ramifications of creating a family through contractual parenting can provide infertile individuals with information that can facilitate informed decisions about their options (Ciccarelli, 1997) and suggestions for improving the surrogacy process for all parties involved.

Examination of two online databases, Psych. Info. and Digital Dissertations (i.e., Dissertation Abstracts), identified only 27 empirical studies (published articles, books, chapters, or doctoral dissertations), from January 1983 to December 2003, that directly studied characteristics and interaction patterns of surrogate mothers; characteristics and interaction patterns of the intended/social parents; and/or attitudes about contractual parenting, surrogate mothers, and intended/social parents (see Table 1).

The research literature primarily describes the motivations and characteristics of surrogate mothers. Many (e.g., Blyth, 1994; Ciccarelli, 1997; Hohman & Hagan, 2001; Migdal, 1989; Preisinger, 1998; Ragone, 1996; and Roher, 1988) are small sample studies of less than 30 surrogate mothers (range of 4 to 28) that primarily analyze qualitative data. A few small studies (Einwohner, 1989; Fischer & Gillman, 1991; Hanafin, 1984; Parker, 1983) assess personality characteristics of surrogate mothers using standardized personality tests. Four studies (Blyth, 1995; Hughes, 1990; Kleinpeter, 2002; Ragone, 1996) examine characteristics or interaction patterns of the intended/social parents and another seven investigate attitudes toward contractual parenting. Finally, we could find only four studies which included comparison or control groups. In three, (Fischer & Gillman, 1991; Hanafin, 1984; Resnick, 1990) surrogate mothers were compared to non-surrogate mothers. The fourth (Hughes, 1990) examined the psychological characteristics of a sample of 95 participants that included both individuals who had become a parent though contracting with a surrogate mother and individuals who had adopted a child.

Below we integrate research on contractual parenting from a number of major subareas. Although it is possible to dismiss this research as preliminary as well

Table 1. Studies on Psychological Aspects of Surrogacy

Studies of the Characteristics and Interaction Patterns of Surrogate Mothers

Author(s)	Source*	Sample	Data Collection Methods	Variables
Baslington, 2002	J	19 surrogate mothers; 6 husbands of surrogate mothers	interviews	relinquishment of the child; psychological detachment process
Blyth, 1994	J	19 surrogate mothers in Great Britain	semi-structured interview	motivations; contact/relationship with intended parents; experience of surrogacy arrangement
Ciccarelli, 1997	D	14 Caucasian women who were surrogates 3–10 years previously (7 IA and 7 IVF surrogates)	open-ended interviews	motivations; relationship with the couple; experience of the surrogacy arrangement; expectations and whether they are met; post birth experiences; satisfaction
Derouen, 1992	D	33 women from one program (21 religious, 12 not religious)	telephone interview and survey	motivations; religiosity
Eimwohner, 1989	BC	50 women who volunteered to be surrogates	semistructured interview; projective and non-projective personality tests	motivations; personality characteristics
Fischer & Gillman, 1991	J, based on D	42 pregnant women (21 involved in surrogate programs across the U.S. and 21 not involved; in each group 20 Caucasian and 1 Hispanic)	quantitative questionnaires	level and quality of attachment; attitudes toward pregnancy; social support (Personal Resources Questionnaire)
Hanafin, 1984	D	21 surrogate mothers not yet in final 2 months of pregnancy and 21 comparison group mothers (20 Caucasians and 1 Hispanic in each group)	Questionnaires; open-ended interview; personality inventory	motivations; personality characteristics; feelings during pregnancy
Hohman & Hagan, 2001	J	17 surrogate mothers from one program, most of whom had given birth 5–7 years previously (13 White, 4 Hispanic)	semi-structured interview	experiences and satisfaction with process; relationship with couple

(Continued)

Table 1. (*Continued*)

Author(s)	Source*	Sample	Data Collection Methods	Variables
Kleinpeter & Hohman, 2000	J	15 women in a California surrogacy program (13 White, 3 Hispanic, 1 other)	personality inventory (NEO-R)	neuroticism; extroversion; openness; agreeableness; conscientiousness
Migdal, 1989*	D	9 women from a surrogate mother program	open-ended interview	motivations; relationship with couple; post-birth experiences; relinquishing the infant
Parker, 1983	J	125 White women who applied to be surrogate mothers	interview	motivations; demographic characteristics; pregnancy/abortion history
Preisinger, 1998	D	4 surrogate mothers	open-ended, in-person interview; telephone interview	experiences as a surrogate; relinquishing the child
Ragone, 1996	J, based on D	28 predominantly White women at 6 different programs	ethnographic; 28 formal interviews plus conversations and observation of program activities	motivations; interaction and relationship with the couple; gender roles
Resnick, 1990	D	43 surrogate mothers and 34 control women	Questionnaire; personality inventory (MMPI subscales)	attachment history; nurturance; relinquishing the child
Roher, 1988	D	13 women (interviewed) and 157 surrogates' files at one program	interviews, files	social and reproductive roles
van den Akker, 2003	J	24 surrogate mothers (11 IFV and 13 AI surrogates)	semi-structured interviews; standardized questionnaires	motivations; experiences; support concerns; disclosure and relinquishment issues; quality of life; psychopathology

Studies of the Characteristics and Interaction Patterns of the Intended/Social Parents

Blyth, 1995	J	20 intended/social parents (9 married couples, 1 women, 1 man) who were members of a British self-help group	interviews	decision-making about surrogacy; relationship with surrogate; reactions of others; beliefs about telling child about genetic origins; gender relationships in surrogacy arrangements
Hughes, 1990	D	95 Caucasian individuals, including 39 couples in three groups: surrogacy, private adoption agency, or independent adoption through an attorney; also a comparison group of 20 parents of preschool children	self administered mailed questionnaire	sensation seeking, self esteem, gender role behaviors; locus of control; social desirability; influences on decisions about whether to participate in 6 different methods of assisted parenthood; demographic characteristics
Kleinpeter, 2002	J	26 parents (24 mothers, 2 fathers)	qualitative methods; telephone interview	decision-making; support; relationship with their surrogate
Ragone, 1996 (note that this study is also mentioned above)	J, based on D	clients and staff of 6 surrogacy programs	ethnographic; observation of consultations with prospective couples and other program activities	relationship with surrogate; intended mother=s bond with surrogate; intended mother=s experiences

(Continued)

Table 1. *(Continued)*

Studies of Attitudes Toward Surrogacy Arrangements

Author(s)	Source*	Sample	Data Collection Methods	Variables
Dunn, Ryan, & O'Brien, 1988	J	485 White and 248 African American undergraduate college students in Southeastern U.S.	questionnaire	attitudes toward 6 methods for dealing with infertility including surrogate motherhood
Grand, 1997	D	115 females and 38 males (72 infertile, 81 non-infertile); 61% Hispanic, 21% White	structured questionnaire	attitudes and opinions toward methods of dealing with infertility including surrogacy
Holbrook, 1996	J	300 social workers, 71% female and 91% White	47-item mail questionnaire	views of rights of participants involved in surrogacy; other ARTs and adoption
Krishnan, 1994	J	5,315 Canadian women (aged 18–49)	1984 national fertility survey using telephone interview	attitudes toward commercial surrogacy; surrogate mothers and other ARTs; sociodemographic characteristics
Lasker & Borg, 1994	B	1) over 200 persons who were infertile most contacted through support groups plus persons connected in some way to infertility clinics and surrogacy programs; 2) 165 mostly White and middle class students at 2 colleges in Pennsylvania	1) taped in-person and phone interviews; questionnaire; 2) survey	1) philosophies of surrogacy programs; relationship between surrogate mother and couple; other ARTs; trauma of infertility; 2) attitudes toward surrogacy and other ARTs
Miall, 1989	J	71 involuntarily childless women (aged 24–45, white, middle class)	survey	Attitudes toward surrogacy; AI and adoption
van den Akker, 2001	J	42 women attending infertility clinics in Great Britain (aged 25–45)	retrospective questionnaire	willingness to disclose mode of starting a family through surrogacy; other ARTs and adoption; acceptability of each of these methods

Note. *Sources are: J = journal, D = dissertation, BC = book chapter; B = book.

as identify significant methodological flaws in many studies, the consistency of results often is impressive. Moreover, empirical data offer little support for widely expressed concerns about contractual parenting being emotionally damaging or exploitative for surrogate mothers, children or intended/social parents.

Attitudes About Surrogacy

A reproductive technology will be used only if it is considered acceptable by potential consumers. Studies to date support the assertion that contractual parenting, especially when it involves a financial payment to the birth mother for carrying a child, is perceived as the least acceptable of all assisted reproductive technologies, with approval percentages ranging from below 10% to about 25% in surveys of college students (Dunn, Ryan, & O'Brien, 1988; Lasker & Borg, 1994), *Psychology Today* readers (cited in Lasker & Borg, 1994, p. 168), Canadian women of child-bearing age (Krishnan, 1994), and infertile women in Great Britain (van den Akker, 2001). This is a much lower percentage than people who approve of or state that they might consider IVF, embryo transplant, and AI by husband (Dunn et al., 1988). In general, methods that involved third parties (AI by donor and surrogacy) have lower approval rates.

Demographic differences in approval rates appear quite minimal. In Krishnan's (1994) analysis of data from a Canadian national fertility survey of over 5,000 women in the childbearing years, size of family of origin, age, and religiosity were negatively associated with approval of commercial surrogacy whereas education was positively associated. Together, however, these and other demographic variables explained only seven percent of the variance in attitudes toward commercial surrogacy. One characteristic that may be associated with approval of contractual parenting is infertility itself. Miall (1989) found that 73% of a small sample of women diagnosed as infertile in Ontario, Canada stated they approved in principle of surrogate motherhood. In the larger Canadian fertility survey, childless women had the most favorable attitudes toward contractual parenting. However, differences in attitudes between women known to be sterile and fecund women were very small (Krishnan, 1994). Thus, it is unclear if an inability to produce a child of one's own leads to greater acceptance of surrogacy, as an unwelcome but necessary reproductive option.

Surrogate Mothers

Characteristics and Motivation

There has been great curiosity about what the typical surrogate mother is like. While it is easy to understand the unhappiness and despair that motivate an infertile, childless couple, who desire children, to enter into a surrogacy arrangement, the motives of women who choose to be surrogate mothers, despite general public disapproval of third party assisted reproduction, are more puzzling and more suspect. Contrary to popular beliefs about money as a prime motive, surrogate mothers overwhelmingly report that they choose to bear children for others primarily out of altruistic concerns (Ciccarelli, 1997; Hanafin, 1984; van den Akker, 2003). Although financial reasons may be present, only a handful of women mention money as their main motivator (e.g., Hanafin, 1984; Hohman & Hagan, 2001; Migdal, 1989; for exceptions see Einwohner, 1989, in which 40% of women state the fee was their main, although not their only, motivator and Baslington, 2002, in which 21% only mentioned money as a motivator). Rather, the women have empathy for childless couples and want to help others experience the great joy of parenthood. Also, some want to take a special action and, thereby, gain a sense of achievement (Blyth, 1994; Ciccarelli, 1997, Hanafin, 1984) or enhance their self-esteem (van den Akker, 2003).

Some surrogate mothers report enjoyment of pregnancy as a motive. In addition, a substantial minority of women have experienced a prior loss, such as an abortion or having given up a child for adoption that they perceive as motivating them to be a surrogate (Parker, 1983). Interestingly, Parker reported 26% of his sample of women seeking to be surrogate mothers previously had a voluntary abortion and 9% previously placed a child up for adoption. However, we could not find documented evidence to suggest that these events are more prevalent for surrogate mothers than other birth mothers with similar demographic characteristics.

It is possible that verbal self reports reflect socially accepted reasons rather than underlying motivation. Ragone (1994) commented that the "stated motivations of surrogates are often expressed in what can be described as a scripted manner" (p. 52) of consistency and conformity in surrogate responses. Based on her ethnographic research at six surrogacy centers including interviews with 28 surrogate mothers, Ragone (1994, 1996) contends surrogate mothers report motivations that reflect traditional culturally accepted ideas about reproduction, motherhood, and family while devaluing characteristics of the surrogacy relationships, such as financial payment, that depart from traditional values and beliefs. Although they may value traditional motherhood, surrogate mothers are engaging in a behavior that represents a radical departure from traditional views of motherhood and family. Ragone believes that many women become surrogate mothers in order to transcend the limits of traditional female roles by doing something special for another couple while at the same time they struggle to confirm the value of such roles.

The literature also provides information about the sociodemographic characteristics and personal traits of women who become surrogate mothers. Scholarly discussions of social class and socioeconomic issues have deplored the potential for exploitation of poor women as surrogate mothers (e.g., Tangri & Kahn, 1993; Ciccarelli, 1997). It is often implied that surrogacy contracts could exploit poor, young, single, or ethnic minority women (Ciccarelli, 1997). Yet, the data do not support this since, in fact, most surrogate mothers are in their twenties or thirties, White, Christian, married, and have children of their own (Baslington, 2002; Ciccarelli, 1997; Kleinpeter & Hohman, 2000; Ragone, 1996; van den Akker, 2003). However, our discussions with surrogacy agencies and professionals (e.g., Center for Surrogate Parenting, H. Hanafin, personal communication, November 12, 1997) suggest that it is likely that surrogate demographics are due, at least in part, to the screening which is utilized by surrogacy agencies in selecting candidates to be surrogates. These screening procedures are specifically designed to circumvent arguments that the process could be exploitive of poor, young, ethnic women (Ciccarelli, 1997).

Surrogate mothers' family incomes are most often modest (as opposed to low), and they are from working class backgrounds. Also, as previously stated, most do not report financial considerations as their main motivation for being surrogates (Ciccarelli, 1997). Moreover, women of color are greatly underrepresented among surrogate mothers (Ciccarelli, 1997). Despite lack of research support for the economic exploitation of surrogate mothers, it is understandable how some scholars would be concerned that the disparities in income and social class between surrogate mothers and intended parents could create the potential for exploitation.

Personality traits of surrogate mothers also are of interest. Are these women mentally stable with personality traits in the normative range or do they have dysfunctional characteristics? Small, non-representative samples; lack of control groups; and ambiguous or flawed comparisons with test norms make it difficult to reach any conclusions about the personal traits of women who become surrogate mothers. At best, it cautiously can be stated that most surrogate mothers are within the *normal* range on personality tests such as the MMPI (Einwohner, 1989; Kleinpeter & Hohman, 2000; van den Akker, 2003). Moreover, they do not differ from mothers who are not surrogate mothers in reported early attachment history (Resnick, 1990). On the other hand, women willing to be surrogates may be more independent thinkers (Migdal, 1989), less bound by traditional moral values. Kleinpeter and Hohman (2000) report that surrogate mothers scored lower on Conscientiousness and Dutifulness on the NEO Five Factor Test, which could suggest that they have a more flexible approach to the application of moral and ethical principles as currently defined by traditional values about family and the meaning of motherhood.

Experienced Satisfaction

Surrogate mothers generally report being quite satisfied with their experiences as surrogates. Ciccarelli's (1997) research was a follow-up study in which 14 participants (7 traditional surrogates and 7 gestational surrogates) were interviewed 5 to 10 years after serving as surrogate mothers. The surrogates were identified through surrogacy agencies with which the surrogates had worked, and were selected based on their willingness to voluntarily participate in the study. Nearly all participants were California residents, Caucasian, and in their 20s or 30s; most were Christian and had at least one child prior to functioning as a surrogate. All were satisfied with their decision to become a surrogate and perceived the experience as enriching (Ciccarelli, 1997). Nevertheless, pre- and post-birth experiences, relationship with the contracting couple, and whether expectations about surrogacy are met are important influences on the surrogate mothers' level of satisfaction (Ciccarelli, 1997). Several studies confirm that the surrogate mother generally forms a relationship with the couple rather than the child (Baslington, 2002; Ciccarelli, 1997; Hohman & Hagan, 2001; Ragone, 1996). Women consistently refer to the developing fetus as the couple's child, rather than their own (Ciccarelli, 1997), and they evidence lower attachment to the fetus during pregnancy than other pregnant women (Fischer & Gillman, 1991). Thus, it is the quality of the relationship with the couple that largely determines the surrogate mother's satisfaction with her experience (Baslington, 2002; Ciccarelli, 1997; Hohman & Hagan, 2001). Moreover, further examination shows that the relationship with the couple is primarily a relationship with the intended mother (Blyth, 1994; Hohman & Hagan, 2001; Ragone, 1996). In effect, the pregnancy is defined as a woman's role and the two women share experiences and events related to the pregnancy, thus often forming a close bond.

Unmet expectations are associated with dissatisfaction with the surrogacy experience. In Ciccarelli's (1997) study, 4 of 14 women had unmet expectations and, in two of these cases, expectations regarding level of closeness with the couple were not met. Such unmet expectations can arise at any time during the initial surrogacy arrangements, pregnancy, or many years post birth (Ciccarelli, 1997). Couple interaction with the surrogate immediately post birth appears important. If the surrogate mother is allowed to see and hold the baby and she feels she is being treated with respect, her satisfaction level is high (Hohman & Hagan, 2001).

Few studies have examined surrogate mothers' relationship with the couple and satisfaction levels up to 10 years after the birth of the child. (Ciccarelli, 1997; Hohman & Hagan, 2001). Most surrogate mothers have some limited contact with the social parents (e.g., pictures of the child, telephone calls) for several years after the birth. Long-term satisfaction continues to depend on the surrogate mother's relationship with the couple and whether her expectations about the relationship and types of contact with the couple and child are met. According to Ciccarelli (1997), as contact with the couple begins to taper off, a minority of surrogate mothers become increasingly dissatisfied with the surrogacy arrangement. The type of surrogacy does not in itself seem to influence satisfaction, rather, the perception of the surrogate regarding her relationship with, and importance to, the couple is determinative (Ciccarelli, 1997). It is particularly damaging if the surrogate mother begins to feel increasingly abandoned by the couple over time (Ciccarelli, 1997).

Effects on Other Social Relationships

Almost all surrogate mothers identified in the literature have a child or children of their own, and the majority are married or with a partner (Baslington, 2002; Ciccarelli, 1997). Although family disapproval is not absent entirely (van den Akker, 2001), surrogate mothers perceived their decision to bear a child for a couple as having a positive effect on close family members, in particular their children (Ciccarelli, 1997), or at worst perceive their own children as not being negatively impacted by the experience (Hohman & Hagan, 2001). Half of the women in Ciccarelli's (1997) study reported becoming closer to a family member as the result of the surrogacy experience and nearly three-quarter of the surrogates indicated that the experience affected their own children in a positive way.

Husbands and partners in the Hohman & Hagan (2001) study were generally seen as supportive of surrogacy. Most women who did not have partners reported some support from close family members, friends, the couple, and/or the surrogacy agency director (Ciccarelli, 1997). In contrast, extended families and friends showed mixed reactions. Less than one-third of the responses by extended family were consistently supportive. In Ciccarelli's (1997) research more than half of the participants experienced conflict in interpersonal relationships as the result of being a surrogate mother and over 40% mentioned having lost a relationship as a result.

Negative Effects

Thus far, we have painted a generally rosy picture of the outcomes of surrogacy arrangements for the birth mother. Nevertheless, navigating this rocky terrain in which few known ground rules exist is not easy and may have significant negative emotional effects for some surrogate mothers (Baslington, 2000; Ciccarelli, 1997). Mild and transient negative repercussions of the surrogacy experience probably occur in varying degrees for all women. Most are general side effects of pregnancy that involve physical discomfort, experienced by all birth mothers. Women who become surrogate mothers usually have good reason to believe they will have normal, relatively easy pregnancies, but all experience routine aches and pains and some experience complications that may lead to a difficult pregnancy (Ciccarelli, 1997).

Occasionally women regret their decision to become a surrogate (Blyth, 1994; Ciccarelli, 1997). As previously stated, dissatisfaction with the surrogacy arrangement may increase over time as contact with the couple diminishes (Ciccarelli, 1997). Blyth (1994) identified 2 out of 17 women who regretted their decision. His is also the only study that reports a significant minority of women (about 25%) who experienced significant emotional distress in giving up the child. It is unclear whether the dissatisfaction stems from the surrogacy process itself, the lack of therapeutic intervention, or both. The considerable proportion of emotionally distressed and dissatisfied women may be exacerbated by the lack of professional support for women in Great Britain, where surrogacy agencies are illegal. However, surrogacy arrangements, including those involving payment to the surrogate mother, are not banned.

Professional support and intervention, including therapy, before and during the surrogacy process may maximize satisfaction rates among surrogates (Ciccarelli, 1997). In addition to initial screening of potential surrogates, most surrogacy agencies offer psychological support and intervention throughout the entire process (Ciccarelli, 1997). Nearly all surrogate mothers in Ciccarelli's research indicated that their satisfaction was increased due to access to competent professionals who helped guide them through the process and deal with emotional issues and any problems that arose. This raises the question of whether the therapeutic process alters one's inherent reaction of experiencing emotional distress when participating as a surrogate mother. This may explain, in part, why the incidence of dissatisfaction increases over time when there is no longer active participation in therapy by the surrogate mother (Ciccarelli, 1997). In contrast to the Ciccarelli (1997) study, another study (van den Akker, 2001) indicated that the perceived usefulness of counseling varied among surrogates. Of the 15 surrogates who participated in this study, 1 indicated that she received "a lot" of practical support, 7 received "some" practical support, and 7 received "no" practical support from counselors (van den Akker, 2001). None of the women indicated that they received "a lot" of emotional support, 5 received "some" emotional support, and 10 received "no" emotional support from counselors (van den Akker, 2001). Since there are no data on how often therapy is needed and for what specific reasons, this may be an important area for future research.

In an effort to reduce negative effects, many surrogacy agencies in the United States will contract with only women who have previously given birth and have children of their own. This maximizes chances of a successful birth and fulfillment of the surrogacy contract; women who have experienced bonding with a child during pregnancy may have a more realistic perception about what it will be like to relinquish a baby to another couple (Ciccarelli, 1997). Additionally, the negative effects reported in Blyth's study (1994) may be due, in part, to the fact that all but two of the surrogate mothers were traditional surrogates. In van den Akker's (2001) study, all the genetic (i.e., traditional) surrogates reported believing a genetic link to the child was unimportant while most of the gestational surrogates disagreed. This raises the question of whether surrogates select the type of surrogacy that fits with their beliefs and values. These types of issues are routinely addressed by surrogacy professionals during the screening process. The above evidence supports the importance, as many surrogates themselves have noted, of using a competent agency that includes a mental health professional in order to minimize potential psychological problems and other negative effects of the surrogacy process (Ciccarelli, 1997).

The Intended/Social Parents

The large bulk of psychosocial evidence on contractual parenting is based on interviews with traditional surrogate and gestational surrogate mothers. We identified only four studies that included intended/social parents. Blyth (1995) interviewed 20 individuals (9 couples, 1 man and 1 woman) in Great Britain who had a child through surrogacy or were in earlier phases of surrogacy arrangements. Participants were recruited through a self-help group for intended parents and surrogate mothers. The majority of couples contracted with traditional surrogates. In all but one case, the decision to consider surrogacy was made by the wife alone who then convinced her husband to consider surrogacy (Blyth, 1995). In general, the accounts of intended/social parents mentioned the difficulties and anticipated embarrassment in finding out information about the potential surrogate mother, and providing her with information about themselves. Also, some noted the awkwardness of maintaining contact with the surrogate, especially for the father, presumably because of the ambiguity of gender relationships in surrogacy arrangements (Blyth, 1995). Responses of others were reported as generally positive to the arrangement, although usually only close family members and friends had been told.

Kleinpeter (2002) used grounded theory to examine telephone interview data from 26 parents (24 women) involved in surrogacy arrangements through one California-based surrogacy program. Most intended/social parents were married, white, and had incomes over $80,000 per year. One dominant theme that emerged was the desire to have a genetic link to the child. Although all parents had concerns about the surrogacy arrangements (e.g., financial stress, legal issues, concern that surrogate would not take care of herself and the unborn child), most described their relationship with the surrogate during the pregnancy as positive. Areas of conflict that sometimes emerged primarily related to the surrogate not attending to the health of the fetus. Close to half of the participants perceived their families (mainly parents ad parents in-laws) as supportive while many others experienced mixed reactions; in contrast, almost all described friends as supportive.

Ragone's (1996) wide ranging ethnographic study of six surrogate programs included an analysis of couples. Although not formally interviewed, an unspecified number of couples were observed interacting with program directors and being interviewed during consultation with a staff member. Ragone (1996) concluded that biological relatedness was a primary motivation for couples' deciding to pursue surrogacy. However, surrogacy violated accepted cultural norms, thus requiring couples to use various cognitive dissonance reduction strategies to resolve the problems and ambiguities associated with surrogate parenthood. In particular, in

AI surrogacy, the father feels discomfort and awkwardness that a woman other than his wife is the mother of the child (Ragone, 1996). Two primary strategies employed by the couple and the surrogate mother to resolve cognitive dissonance are to (a) de-emphasize the man's role by defining pregnancy and birth as women's business; and (b) downplay the significance of the biological link to the child (Ragone, 1996). The intended mother often justifies the lack of genetic ties to the child through development of a mythic conception of the child that emphasizes her intentionality in the process (it is her desire that ultimately brings the child into being; Ragone, 1996). Moreover, she develops a relationship with the surrogate mother and experiences pregnancy by proxy (e.g., attending Lamaze classes, being present in the delivery room, going to medical appointments). Thus, reproduction is defined as primarily a woman's concern.

Finally, Hughes (1990) compared the personal characteristics of 53 intended/social parents from a surrogacy program with 42 individuals who adopted children and 20 control subjects. All groups were generally college educated, Caucasian, professional, and had high average self-esteem. Those involved with the surrogacy program were older, had higher household incomes, and were less likely to be Catholic than other participants. In addition, they scored lower on the Marlow Crowne Social Desirability Scale, indicating less need to present in a socially desirable way (Hughes, 1990).

The high socioeconomic status of intended parents is to be expected as the financial costs of surrogacy are high. In addition to the $10,000–20,000 paid to the surrogate mother, the couple must incur many other costs such as payment to the surrogacy agency and all medical expenses leading to a typical total cost of between $25,000 and $100,000, with IVF surrogacy on the high end (Center for Surrogate Parenting, 2003). All studies found that intended/social parents are well off financially; for instance, Ragone (1996) found an average income of over $100,000 for contracting couples. Thus, except in rare cases of non-commercial surrogacy usually for family members or friends who cannot have a child, contractual parenting is possible only for the wealthy or upper middle class. The lack of access to surrogacy arrangements for lower income infertile couples is a major ethical and sociopolitical concern for feminists and others who support equal access to reproductive health services for all individuals regardless of socioeconomic status or racial/ethnic origins.

Children Resulting from Contractual Parenting

We could find no studies examining the cognitive or social development of children born as the result of surrogacy. An exploration of related areas revealed that there are no appropriate parallels. Adoption does not appear to be a good comparison because adopted children have no genetic connection to either parent and adoption is a more socially acceptable action that does not violate traditional norms.

There are some studies that may provide some limited comparison. Research on the cognitive and social development of children produced through other assisted reproductive technologies, most usually IVF, may be tangentially related, while studies of children conceived through egg donation provide a somewhat better comparison. Reviews of the literature suggest that IVF children in developmental stages from infancy through adolescence show comparable cognitive functioning to other children and in some cases score higher in social and communication skills (McMahon, Ungerer, Beaupaire, Tennant et al., 1995; Van Balen, 1998). Some studies even suggest that the experience of infertility and use of Assisted Reproductive Technologies (ARTs) actually may be beneficial for parent-child relationships (Gibson, Ungerer, McMahon, Leslie, & Saunders, 2000; Hahn & DiPietro, 2001; VanBalen, 1996). One study (Golombok, Murray, Brinsden, & Abdalla, 1999) comparing egg donation, donor insemination, adoptive families, and IVF families reported no overall differences among groups in quality of parenting or psychological adjustment of children aged three and a half to eight. It seems likely that, from the child's perspective, the mechanisms of how a pregnancy was achieved would be a minimal psychological issue compared to whether one's birth mother chose not to keep the child. Research to date is only suggestive and, clearly, it is necessary to explore the social, psychological, and cognitive development of children born through surrogacy.

Notwithstanding the foregoing, one underlying issue for all types of ARTs, but especially those that involve third parties, is whether, when and what to tell the child about his or her origins. Blyth reported that all intended parents in his study believed the child should eventually be told the truth about his or her biological origins (Blyth, 1995). However, there is no consensus due to a lack of research on this issue.

Future Directions

Research Issues

There is an abundance of potential research questions involving contractual parenting that appears worthy of investigation. Both researchers and those debating the moral, ethical, legal, and social aspects of contractual parenting have supported the need for more empirical data and proposed questions of interest. While it is not difficult to identify research directions, it is more challenging to prioritize directions. In this section we describe several research questions that warrant priority.

Clearly, a primary focus should be on the potential impact on the children that are born as a result of third party assisted reproduction as well as children in the surrogate's family. Although there is no particular reason to believe that AI and IVF children born as a result of surrogacy arrangements will differ in development from other children born through ARTs, studies of the development of the offspring of surrogacy arrangements still are important. Pragmatic issues provide guidance for future research on the post-birth effects of surrogacy arrangements. According to Blyth (1995), many social parents intend to tell their child about his or her origins. As far as is known, however, few children have been informed presumably because of their still-young age. If, indeed, interpersonal issues are more important for the child's development and well-being than the fact that conception occurred through assisted reproduction, then researchers need to consider questions such as how best to explain their origins and the birth mother's relationship to children of various ages, how much contact should the birth mother have with the child, and do different issues arise for children born through traditional versus gestational surrogacy. Research issues involving communication with the child include when—or if—to tell children of their biological origins, how much to reveal, and the long-term consequences of deception versus honesty. Issues related to birth mother contact with the child that need investigation involve the benefits or detriments of the child remaining in contact with the surrogate mother and the long-term impact on the family dynamics—both for the intended parents and the surrogate and/or her family—in cases where all parties stay in contact as well as cases where contact diminishes or stops. In some cases, critical analysis of extant parallel bodies of research on, for instance, other types of assisted reproduction or adoption may be most appropriate.

Another priority is to heighten access to participant populations and enhance their voluntary response rates to research requests. Both surrogate mothers and intended/social parents have a vested interest in promoting the view that surrogacy is acceptable and that those who commit to surrogacy contracts are well-adjusted individuals. In addition, all parties are interested in the cognitive and social development and best interests of the resultant child. Moreover, parties to surrogacy agreements may be motivated to support extensions of this option to other infertile couples who desire a family and to increase public understanding of this issue. These are powerful hooks that can be used to interest these parties in voluntarily participating in research. Of course, identification and recruitment of samples of surrogate mothers and intended parents is not easy. Most often such identification has occurred through surrogacy agencies or support groups. As access to the Internet increases and many surrogates and commissioning couples use net-based resources to attempt to find a match, this, too, may prove a valuable recruitment avenue.

The issue of what to research is largely defined by studies that are strikingly absent. More attention has been given to the surrogate mother than to the intended parents. Moreover, although there is research on relationships of the surrogate and the intended parents and their perceptions of their social networks, these studies (with the possible exception of Hohman & Hagan, 2001) are not based on a firm conceptual or theoretical framework about complex interpersonal relationships under conditions of stress. Yet, surrogacy arrangements involve complex interpersonal processes and interactions. There are three individuals, all with their own needs and desires, plus their families, which, in the case of the surrogate, usually include children who are minors.

Although we do not advocate studies of the motives or personalities of women who choose to become surrogates as a priority, another post-birth effect that needs more attention is the potential level of regret experienced by surrogate mothers over time. In particular, we need to determine how psychological intervention alters perceived dissatisfaction with the surrogacy process, for instance, by comparing the level of satisfaction of the surrogacy process of surrogate mothers who receive different types or amounts of counseling both before entering into surrogacy contracts and during the surrogacy process.

Finally, the future of surrogacy arrangements is dependent on what people find acceptable both personally and as a matter of public policy. In part, surrogacy has not evoked as much controversy as abortion because it is relatively rare. Still, it touches upon basic beliefs about what constitutes parenthood, the importance of a genetic link to the child, and gender relationships. World views and values regarding family and gender roles of anti- and pro-surrogacy groups should be studied as should differences in the positions of pro- and anti-surrogacy feminists. Also, it would be useful to analyze the basic cultural values that have led countries such as Australia to outlaw surrogacy. Such studies of cultural beliefs, values, and attitudes will provide more valuable information than have previous surveys that simply determine the percentage of a group supportive of a specific type of surrogacy arrangement.

Treatment Service Issues

Because of the deficit of empirical evidence, it is premature to advocate many specific changes in treatment services or social policy. There are general approaches, however, that should be followed to alleviate some of the anxiety, distress, and post-birth regret experienced by one or more of the parties involved. For instance, it tentatively can be assumed that satisfaction with contractual parenting is largely influenced by satisfaction with the relationship between the surrogate and the commissioning couple, which in turn is largely determined by the extent to which expectations about this relationship are met (Ciccarelli, 1997). Therefore, counselors need to provide accurate information to participants about all phases of the surrogacy process and determine during screening that the parties have adequate personal resources and support networks to withstand the stress and disapproval that engaging in this process may engender. Moreover, it is important that counselors and other mental health professionals with knowledge of the potential pitfalls of surrogacy arrangements be available to participants at all stages (pre-contract, during pregnancy, post-birth, and long term).

Legal and Public Policy Issues

Surrogacy as a process can "go bad" at many points. Although this souring of relationships and resultant high profile legal cases are relatively rare, statutes that require use of reputable surrogacy agencies with well-trained mental health and legal professionals can minimize both the contractual disasters and the milder, but still painful, long-term feelings of regret of some birth mothers. Couples who choose this option usually have exhausted more traditional alternatives, and have lived with the stress of infertility for years. As elaborated in Ciccarelli and Ciccarelli (this issue), the ambiguity of the legal situation in some jurisdictions makes it most difficult to assuage the additional stress that intended parents experience because of the myriad of things that could go wrong in their relationship with the surrogate. Any statutes that clarify the procedures and allow for pre-birth adoption of the baby can help alleviate the anxiety evoked by the uncertainty and ambiguities of surrogacy arrangements for commissioning couples, but perhaps at the cost of the rights of the birth mother.

Finally, both acceptability and accessibility will determine the extent to which this new technology is used. To the extent that public policy *institutionalizes* this option, it will become more acceptable to couples with no other options and to women motivated to perform an altruistic service. There will always be cultural groups, however, who because of basic religious values, will find such arrangements unacceptable or even immoral.

Greater focus on the prevention and early treatment of causes of infertility such as sexually transmitted diseases can reduce the need for surrogacy as well as other expensive ARTs. Yet, contractual parenting appears to be here to stay. Thus, the politics of social class and socioeconomic resources need to remain in the forefront. A remaining predominant issue for third-party assisted reproduction, as well as most other ARTs, is unequal availability, with access usually limited to the top socioeconomic echelon of our society. Unless sweeping changes in the structure of health care occur or disparities in socioeconomic status are reduced, this situation is unlikely to change.

References

Bartholet, E., Draper, E., Resnik, J., & Geller, G. (1994). Rethinking the choice to have children: When, how and whether or not to bear children. *American Behavioral Scientist, 37*, 1058–1073.

Baslington, H. (2002). The social organization of surrogacy: Relinquishing a baby and the role of payment in the psychological detachment process. *Journal of Health Psychology, 7*, 57–71.

Behuniak-Long, S. (1990). Radical conceptions: Reproductive technologies and feminist theories. *Women and Politics, 10*(3), 39–64.

Blyth, E. (1994). "I wanted to be interesting. I wanted to be able to say 'I've done something with my life'": Interviews with surrogate mothers in Britain. *Journal of Reproductive and Infant Psychology, 12,* 189–198.

Blyth, E. (1995). "Not a primrose path": Commissioning parents' experiences of surrogacy arrangements in Britain. *Journal of Reproductive and Infant Psyuchology, 13,* 185–196.

Burr, J. (2002). "Repellent to proper ideas about the procreation of children": Procreation and motherhood in the legal and ethical treatment of the surrogate mother. *Psychology, Evolution and Gender, 2,* 105–117.

Center for Surrogate Parenting. (2003). Retrieved November 13, 2003 from http://www.creatingfamilies.com

Centers for Disease Control and Prevention, National Center for Chronic Disease Prevention and Health Promotion. (n.d.). *1995 assisted reproductive technology success rates: National summary and fertility clinic reports.* Retrieved April 18, 2003, from http://www.cdc.gov/nccdphp/drh/archive/arts/index.htm

Ciccarelli, J. C. (1997). *The surrogate mother: A post-birth follow-up study.* Unpublished Doctoral Dissertation. Los Angeles: California School of Professional Psychology.

Deroven, D. M. (1992). The role of religion in surrogate mothers' motivations. *Dissertation Abstracts International, 53*(6-B), 3142.

Dunn, P. C., Ryan, I. J., & O'Brien, K. (1988). College students' acceptance of adoption and five alternative fertilization techniques. *The Journal of Sex Research, 24,* 282–287.

Einwohner, J. (1989). Who becomes a surrogate: Personality characteristics. In J. Offerman-Zuckerberg (Ed.), *Gender in transition: A new frontier* (pp. 123–132). New York: Plenum.

Fischer, S., and Gillman, I. (1991). Surrogate motherhood: Attachment, attitudes and social support. *Psychiatry, 54,* 13–20.

Gallagher, M. (1989). Enemies of Eros. In re Marriage of Moschetta (Cal. App. 4th 1218, 1994).

Gibson, R. (1994). Contract motherhood: Social practice in social contexts. In A. M. Jaggar (Ed.), *Living with contradictions: Controversies in feminist social values* (pp. 402–419). Boulder, CO: Westview.

Gibson, F. L., Ungerer, J. A., McMahon, C. A., Leslie, G., & Saunders, D. M. (2000). The mother-child relationship following in vitro fertilization (IVF): Infant attachment, responsivity, and maternal sensitivity. *Journal of Child Psychology and Psychiatry and Allied Disciplines, 42,* 1015–1023.

Golombok, S., Murray, C., Brinsden, P., & Abdalla, H. (1999). Social versus biological parenting: Family functioning and the socioemotional development of children conceived by egg or sperm donation. *Journal of Child Psychology and Psychiatry and Allied Disciplines, 40,* 519–527.

Grand, C. (1997). New reproductive technologies: An overview of attitudes, opinions, acceptance and their consequences. (Doctoral dissertation, Miami Institute of Psychology, Caribbean Center for Advanced studies.) *Dissertation Abstracts International, B 58/03,* 1223.

Hahn, C., & DiPertro, J. A. (2001). In vitro fertilization and the family: Quality of parenting, family functioning, and child psychosocial adjustment. *Developmental Psychology, 37,* 37–48.

Hammer-Burns, L. H., & Covington, S. C. (1999). Psychology of infertility. In L. Hammer-Burns & S. C. Covington (Eds.), *Infertility counseling* (pp. 3–25). Pearl River, NY: Parthenon.

Hanafin, H. (1984). The surrogate mother: An exploratory study. (Doctoral dissertation, California School of Professional Psychology.) *Dissertation Abstracts International, 45*(10-B), 3335–3336.

Hanafin, H. (1999). Surrogacy and gestational carrier participants. In L. Hammer-Burns & S. C. Covington (Eds.), *Infertility counseling* (pp. 375–388). Pearl River, NY: Parthenon.

Hohman, M. M., & Hagan, C. B. (2001). Satisfaction with surrogate mothering: A relational model. *Journal of Human Behavior in the Social Environment, 4,* 61–84.

Holbrook, S. M. (1996). Social workers' attitudes toward participants' rights in adoption and new reproductive technologies. *Health and Social Work, 21,* 257–266.

Hughes, N. J. (1990). *Some characteristics of couples selecting different methods of assisted parenthood.* Unpublished doctoral dissertation, University of Kansas.

Jaggar, A. M. (Ed.). (1994). *Living with contradictions: Controversies in feminist social values.* Boulder, CO: Westview.

Kleinpeter, C. B. (2002). Surrogacy: The parents' story. *Psychological Reports, 91,* 135–145.

Kleinpeter, C. G., & Hohman, M. A. (2000). Surrogate motherhood: Personality traits and satisfaction with service providers. *Psychological Reports, 87,* 957–970.

Krishnan, V. (1994). Attitudes toward surrogate motherhood in Canada. *Health Care for Women International, 15,* 333–357.

Lasker, J. N., & Borg, S. (1994). *In search of parenthood: Coping with infertility and high-tech conception.* Philadelphia: Temple University Press.

Luker, K. (1984). *Abortion and the politics of motherhood.* Berkeley, CA: University of California Press.

Macklin, R. (1988). Is there anything wrong with surrogate parenthood: An ethical analysis. *Law, Medicine and Health Care, 16*(1–2), 57–64.

Mahoney, J. (1988). An essay on surrogacy and feminist thought. *Law, Medicine and Health Care, 16,* 81–88.

Matter of Baby M, 537 A.2d 1227 (1988).

McMahon, C. A., Ungerer, J. A., Beaupaire, J., Tennant, C., & Saunders, D. (1995). Psychosocial outcomes for parents and children after in vitro fertilization: A review. *Journal of Reproductive and Infant Psychology, 13,* 1–16.

Miall, C. (1989). Reproductive technology vs. the stigma of involuntary childlessness. *Social Casework, 70,* 43–50.

Migdal, K. L. (1989). *An exploratory study of women's attitudes after completion of a surrogate mother program.* Dissertation Abstracts International, *49*(12-A, Pt 1), 3628–3629.

Parker, P. J. (1983). Motivation of surrogate mothers: Initial findings. *American Journal of Psychiatry, 140,* 117–119.

Preisinger, M. A. (1998). Surrogate mother, A phenomenological naming of who she is: Personal story, mythology and dance. (Doctoral dissertation, Pacifica Graduate Institute.) *Dissertation Abstracts International, 59* (9-B), 5137.

Ragone, H. (1994). *Surrogate motherhood: Conception in the heart.* Boulder, CO: Westview Press.

Ragone, H. (1996). Chasing the blood ties: Surrogate mothers, adoptive mothers and fathers. *American Ethnologist, 23,* 352–365.

Raymond, J. G. (1998). Reproduction, population, technology and rights: North and South. *Women in Action, 2,* 75. Retrieved December 26, 2003, from Gender Watch Database, Proquest Information and Learning Web site: http://Oproquest.umi.com.library.alliant.edu

Resnick, R. F. (1990). Surrogate mothers: The relationship between early attachment and the relinquishing of a child. *Dissertation Abstracts International, 51*(3-B), 1511–1512.

Roher, D. R. (1988). Surrogate motherhood: The nature of a controversial practice. *Dissertation Abstracts International, 49*(4-A), 865.

Rothman, B. K. (1989). On surrogacy: Constructing social policy. In J. Offerman-Zuckerberg (Ed.), *Gender in transition: A new frontier* (pp. 227–233). New York: Plenum.

Tangri, S., & Kahn, J. (1993). Ethical issues in the new reproductive technologies: Perspectives from feminism and the psychology profession. *Professional Psychology: Research and Practice, 24,* 271–280.

Taub, N. (1992). The surrogacy controversy: Making and remaking the family. In D. Nelkin (Ed.), *Controversy: Politics of technical decisions* (3rd Ed.) (pp. 227–233). Thousand Oaks, CA: Sage.

Van Balen, F. (1996). Child-rearing following in vitro fertilization. *Journal of Child Psychology and Psychiatry and Allied Disciplines, 37,* 687–693.

Van Balen, F. (1998). Development of IVF children. *Developmental Review, 18,* 30–46.

van Den Akker, O. (2001). The acceptable face of parenthood: The relative status of biological and cultural interpretations of offspring in infertility treatment. *Psychology, Evolution and Gender, 3,* 137–153.

van Den Akker, O. (2003). Genetic and gestational surrogate mothers' experience of surrogacy. *Journal of Reproductive and Infant Psychology, 21,* 145–161.

JANICE C. CICCARELLI is a Clinical Psychologist in Private Practice in Claremont, California. She is an Adjunct Professor in the Psychology Psy.D. Department at University of LaVerne. In her private practice, Dr. Ciccarelli counsels surrogates and couples involved in third party assisted reproduction.

LINDA J. BECKMAN is a Professor in the Clinical Psychology PhD Program of the California School of Professional Psychology at Alliant International University. She is an Associate Editor of the Psychology of Women Quarterly and author of numerous articles on women's reproductive health.

Journal of Social Issues, Vol. 61, No. 1, 2005, pp. 45–65

Critical Issues in Contraceptive and STI Acceptability Research

Lawrence J. Severy*

University of Florida and Family Health International

Susan Newcomer

National Institute of Child Health and Human Development

We review conceptual issues and theoretical frameworks related to users' accept-ability of new technologies designed to protect reproductive health and prevent unwanted pregnancy. Special attention is given to distinctions among different kinds of users' perspectives regarding acceptability, as well as differentiating ac-ceptability from assessments of the efficacy of innovative methods. Emphasis is also given to the larger context of couple decision-making and cultural variation. We argue that concern for sexual pleasure plays a central role in determining user perspectives regarding new methods. The female condom, contraceptive ring, contraceptive skin patch, microbicides, vaccines, emergency contraception, and PERSONA are discussed within the context of the identified critical issues.

The development of new methods for protection against unwanted pregnancies and/or sexually transmitted infections (STIs) is an exhausting, expensive, complex, and not always successful process. Yet, women and their partners seem vigilant and hopeful of finding methods better than the ones they are currently using, or than the ones that are currently available (Severy & Silver, 1993).

Unprotected sexual behavior is now the single most common route of infection for HIV worldwide (Painter, 2001; UNAIDS, 2002). Despite the presence of highly protective ways in which to prevent infection and highly effective methods for

*Correspondence concerning this article should be addressed to Lawrence J. Severy, P.O. Box 13950, Research Triangle Park, North Carolina, 27709 [e-mail: LSevery@fhi.org].

The authors wish to thank the anonymous reviewers for their excellent insights and suggestions. The conclusions expressed in this article do not necessarily reflect the policies of the University of Florida, Family Health International, or the National Institute of Child Health and Human Development.

preventing unwanted pregnancy, people continue to have unprotected sex, and continue to get infected and/or pregnant when they do not want to (Painter, 2001; UNAIDS, 2002). Consequently, in the past several years, women's health advocates and researchers have detailed (a) the urgent need for the development of female initiated methods that, when applied topically, can prevent the spread of HIV and other STIs (as some couples are not satisfied with current alternatives), as well as (b) the urgent need for more acceptable methods to prevent unintended pregnancy (e.g., Stein, 1996). This plea is exemplified in a 2001 national survey of women and OB/GYNs by the American Medical Women's Association indicating a desire for "methods that simplify women's lives" (*Contraceptive Technology Update*, 2001, p. 13).

An interesting lens through which to consider these issues is provided by Newcomer (2002). Given that the preponderance of HIV infection is now via un-protected sexual behavior, we must seriously consider the motivation for behavior, specifically sexual behavior, that can interfere with protection against HIV and other STIs. She lists three assumptions: pleasure, power, and procreation (in al-phabetical order). And these three may vary across individuals, relationships, time, location, and culture. Perhaps the least attention in our current article is paid to procreation, but the other two are extremely important to our thoughts about ac-ceptability. Procreation is a factor, however. Consider the fact that in some cultures not having children may be as shameful as having HIV, positioning men, women, and couples between two stigmata (Burns, 2003). Some women's choice of birth control methods is designed to preserve their fecundability (Brady, 2003).

Our goals for this article are: to review the complexities involved in under-standing what is meant by "acceptability"; to discuss distinctions between accept-ability and efficacy—and their implications for resultant protection; to present the important attributes of methods leading to acceptability (such as the potential im-pact on sexual intimacy); and, lastly, to discuss—from a behavioral perspective—a number of the more recent alternatives available to those wishing to avoid unwanted pregnancy and STIs.

Acceptability

Our impression is that precious little information exists that can provide es-timates prior to its introduction of how a new method will be integrated into the calculus of conscious choice that individuals are purported to use when deciding upon protection (Davidson & Morrison, 1983). We feel the key question is how best to inform the development and dissemination of effective methods for preg-nancy or disease protection that are *acceptable* to those whose sexual behavior places them at risk. We do not address the issue of whether those at risk, and those not at risk, have the same impressions and values related to acceptability. Our point is that perhaps not all people need protection, hence, the development of such products for these couples would be a waste of time and energy. We intend to focus on those whose sexual behavior places them at risk, those for whom a new method would be of value to the public's health.

We conceptualize acceptability as *the voluntary sustained use of a method in the context of alternatives*. New technology will not be effective if it is not used; and to be used, technology must be designed to fit people's needs rather than vice versa (Keller, 1979). If a product, no matter how effective, is not acceptable—for any number of reasons—it will not be used. While the aim of those who are developing new technologies is to get one that *works*, in order to make it acceptable, one size may not fit all needs. We assume that all contraceptive developers are trying to create methods that are highly effective, but the challenge is more complicated because sustained use will depend upon other behavioral factors involved in acceptability. The *fit* depends upon the context perceived by those involved. Clearly, acceptability of a contraceptive may be variable, depending upon the situation (planned versus spontaneous sex) and on both partners' attitudes towards the product.

As Snowden (1996) argues, most women and men would really rather not be required to use anything while having sex. Hence, Severy and Silver posit that no single method enjoys an overall positive support and that choices are based upon what is believed to be the least offensive or intrusive approach, namely the "least bad alternative" (1993, p. 225).

In addition to the product's effectiveness, its viscosity, color, smell, taste, application method, method of acquisition, required dose for efficacious use, and duration of effectiveness may influence acceptability. Other aspects of the product, such as its actual or perceived effect on sexual intimacy, also have an impact. Severy and Thapa (1994) reframe the issue as one of tolerances. A woman's (or a man's) willingness to *tolerate* various features can be assessed as a preference scale. They demonstrate that tolerance preferences are subject to cultural and subcultural variation. Product characteristics (including side effects) often vary in significance and convey different meanings in different cultures (Good, 1977).

Finally, there are many parties besides the person using the product who can influence adoption and continued use. Decision-making involves two partners—acting in a dynamic fashion and influencing each other's acceptability and behavior. This is not a new idea, as Jackson (an early female British OB/GYN) provocatively argued that the best contraceptive was two reasonable people (Snowden, 1996). Much early research in acceptability and decision-making regarding method choice for family planning and disease prevention depended on data from only one partner—typically the woman (e.g., Darroch & Frost, 1999; Westoff & Ryder, 1977). Not surprisingly, when information from both partners is considered, prediction is improved (Thomson & Hoem, 1998), and it is likely that interventions would be improved as well (Becker, 1996). Theoretical models for understanding method choice and acceptability have been developed and their use in further research could respond to criticisms regarding the relative lack of attention to the couple relationship (Amaro, 1995; Miller, Severy, & Pasta, 2004; Painter, 2001).

Other persons whose views influence acceptability include potential or imagined partners, friends and family, health care providers, health system managers, and manufacturers of products. For example, interviews with women in Mali highlighted the importance of: husbands (opposed), mother-in-laws (opposed), and sisters-in-law (supportive), regarding contraceptive use (Castle, Konate, Ulin, & Martin, 1999). The relative influence of these parties also varies among cultures, as well as among users of different ages and relationship statuses (Hardy, de Padua, Jimenez, & Zaneveld, 1998).

Past research on acceptability typically follows one of three models, appropriate to different stages of product development (Cleland, Hardy, & Taucher, 1990). Hypothetical product acceptability studies determine potential user responses to a product that is not yet developed for use in the target population. Often women are asked to make hypothetical trade-offs among many features of a product, a number of which (such as effectiveness) are unknown at the time of the study. This type of research provides the opportunity for early input from potential users about new methods.

Darroch and Frost (1999) provide a relatively recent example of this work. In a telephone survey of over 1000 U.S. women of reproductive age (18–44), women were asked whether they would be interested in, or would use, vaginal microbicides if they were to be developed and made available. A full 93% said yes—they would use them, if they found themselves in a position where they might be at risk of contracting an STI. The problem is that most women (60%) did not see themselves as vulnerable. Only about 25% were currently in a relationship in which they thought such a product would be useful.

It is unfortunately the case that research asking individuals to evaluate their probable use of a hypothetical method generally has not been successful in predicting actual use of a method. For example, Minnis, Shiboski, and Padian (2003) attempted to assess the relationship between contraceptive acceptability, measured a variety of ways, and actual method use for STI prevention. In their study of barrier methods they distinguished among hypothetical acceptability, product choice following presentation, satisfaction after product use, and sustained use. There were very few significant relationships (i.e., assessments of one domain had little bearing on predictions of the other constructs). They found that both ratings of method satisfaction, and choice from among a set of free samples failed to predict actual use during at least 50% of the opportunities for use. They argue for a reframed approach to assessing acceptability in prevention research so as to effectively predict use.

A second common model is to embed behavioral studies within the clinical trials used to evaluate the safety and efficacy of new products (Severy, 1999).

This approach can provide valuable information about users' actual experience with a method. An example of this approach involves the U.S. trials of a method of contraception found in Europe, known as PERSONA. Researchers (Severy, McNulty, Findley-Klein, & Robinson, 2004) found at least three different groupings of couples entering the clinical trial—including one displaying marital troubles and dislike for their previous forms of contraception, a second experiencing a successful relationship, but desiring a new method due to their dislike of their current choice, and a third group rather satisfied with what they had been using, but looking for something even better. Given these diverse perspectives, it should not be surprising that actual experiences and acceptability varied widely for the three groups during the trial. Although helpful to innovators and developers of new technology, findings from studies married to clinical trials are limited in their capacity to predict probable use of a product in the general population once the product is approved (Elias & Coggins, 2001).

A third model, often referred to as post-marketing research, tracks the adoption and use of a method in actual population settings, studying how the interplay of the individual user, the dyadic relationship, product characteristics, the health system, and the larger social and cultural setting influences use. A demonstrably effective and safe product may find no market, or may encounter other, more subjective, context specific obstacles to its actual adoption and use (cf. IUDs in the United States and India; Treiman, Liskin, Kols, & Rinehart, 1995). This third model offers the advantages of generalization to larger populations, but suffers from the obvious disadvantage that it cannot be completed before an approved product is available. Recent research in Egypt by Tolley, Loza, Kafafi, and Cummings (2003) demonstrates that differential continuation use rates are observed for different methods (see Figure 1). Clearly, long term acceptability might not be as positive as desired—or as positive as first observed.

We have presented some of the elements of an understanding of our sense of the term *acceptability*. For STI prevention methods, or contraceptives, to work effectively, individuals and couples must commit to consistent and correct *use*. We are predominantly interested in sustained use. Simple frequency of use, however, also seems insufficient. Choice and use may not be the same. In some cultures there are no options (except to use nothing—which although a choice is not a real one). Hence, use may or may not reflect the likes and dislikes of those using the method. On the other hand, when we know that alternatives are available, it is easier for us to infer method acceptability from use. We argue for a sense of acceptability demonstrated by long-term use—at least a year or more of consistent and correct use. Yet, we cannot be certain about a product's acceptability unless there are at least two functional methods to choose from. Under such a circumstance, people are not only using a product, they are exercising choice (Elias & Coggins, 2001).

When considering long-term use, it is perhaps appropriate to recall research in a number of health prevention domains related to the concept of "Stages of

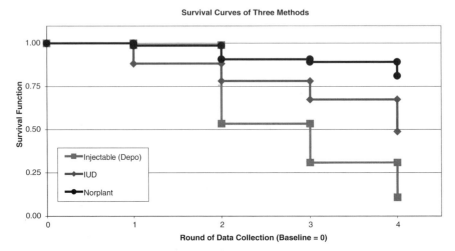

Fig. 1. The percentage of Egyptian women continuing use of their method over time.[1]

[1]From *The impact of menstrual changes on method use: A comparative study among women using the IUD, DMPA and Norplant*, by E. L. Tolley, S. Loza, L. Kafafi, & S. Cummings, 2003, Research Triangle Park, NC: Family Health International. Reprinted with permission of Family Health International.

Change." The basic idea in these theoretical models is that there are stages through which the individual, or couple, must pass if they are to demonstrate consistent long-term adherence to preventative health behaviors. The best known of these is specifically tied to STI prevention and is known as ARRM—the AIDS Risk Reduction Model (Catania, Kegeles, & Coates, 1990). This is a four-stage model including: labeling, commitment, enactment, and maintenance. The last two stages are of most interest to our conceptualization of acceptability. Enactment involves actual behavior designed to reduce high-risk and maintenance involves sustained use over time. In each of these stages, women and their partners experience method use, and they may, or may not, continue use over time.

Other approaches to acceptability can also be linked to the ARRM model. It is easy to view hypothetical acceptability as addressing the first two stages. An individual or a couple would not even think about using prevention methods unless they perceived themselves to be at risk (labeling), and one would not commit to using a method (commitment) unless the features of the method were to their liking. Hence, stages of change reflect stages of acceptability also—ranging from hypothetical to satisfaction with early experience to sustained use over time.

A third conceptual overlay relates to the formal stages of clinical trials for the testing of new products. The four phases include: Phase I—product is tested to be safe and causes no harm; Phase II—product is tested in small targeted populations as to whether it actually works as intended; Phase III—product is tested against a control in large, prospective randomized clinical trials; and Phase IV—success and safety of the product are monitored as it is adopted in the general population. Note that it is possible to assess hypothetical acceptability in conjunction with Phase I, experiential satisfaction can be derived from Phase II and III studies, and sustained use over time in the general population addresses the heart and essence of Phase IV.

Acceptability and Efficacy

One of the more perplexing issues in the development of new technologies involves the distinction and interplay between acceptability and efficacy. Efficacy identifies how well a particular product does its job—if used correctly and consistently. For years people have noted that there is a difference between the theoretical efficacy and the practical efficacy of products. What they are speaking about relates to how well the product *should* work, and how well it *does* work when people use it. This is analogous to physicists' statements about the behavior of entities as studied in a vacuum. Clearly, family planning and STI prevention methods are used in complex human interactions, not in a vacuum. What we have termed "practical efficacy" is also known as the "use-effectiveness" of a method (Elias & Coggins, 2001).

The issue is that government regulatory agencies—in most countries—have very little trouble approving products with high rates of theoretical efficacy. (see Food & Drug Administration (FDA), n.d. *Device Evaluation Information*) Under such circumstances, regulators seem unconcerned with acceptability issues. The assumption seems to be that individuals will *accept* a method containing undesirable textures or unpleasant side effects for a product that works extremely well. This probably happens in a number of cases, especially for unhealthy people. We suspect that a healthy person would be less tolerant of medical interventions that are unpleasant. Only when a product displays marginal efficacy does acceptability data seem to weigh in on the regulators' decision. We argue that regulatory agencies should consider acceptability if they are interested in impacting public health. Users may be willing to *accept* a lower-efficacy product that they can use easily, with enhanced pleasure, or covertly.

As described by Spieler (1997), the level of protection afforded (for example, the number of STI cases averted) depends upon three factors—the efficacy of the method, consistency of use within the partnership (acceptability), and the extent of use in a sub-population. This means that a low efficacy method used with high levels of consistency would offer the same protection as a high efficacy method used less consistently. In fact, a 90% efficacious method used in one-fifth of a person's sex acts, provides less protection than:

- A 75% efficacious method used 30% of the time
- A 50% efficacious method used 40% of the time
- A 25% efficacious method used 80% of the time

It is important to realize that a method with only a 60% efficacy used two-thirds of the time would provide more than twice the protection in comparison with a 90% efficacious method used only 20% of the time (Spieler, 1997). Clearly, regulatory agencies *must* attend to the acceptability of the methods that are presented to them for approval.

There is another level of discourse regarding efficacy and acceptability. Consider, for a moment, a product that might be highly acceptable to the general public, but provides 30% efficacy. In offering a menu of prevention options, the harried health care worker, struggling with a stretched budget, may be unwilling to supply the product. Clearly, efficacy also matters.

Importance of Impact on Intimate Behavior

There are a number of attributes leading to method acceptability (Davidson & Jaccard, 1979; Marshall, 1977; Severy & Silver, 1993). Increasingly, the preponderance of evidence (Albarracin et al., 2001; Ellertson, Elias, & Coggins, 1999; Padian, 1999) seems to indicate that a woman's acceptance of any form of protection from an unwanted pregnancy or STI, and her partner's acceptance, is very much a function of their beliefs that using the method would have a meaningful impact on sexual intimacy. We assume that one of the main reasons people have sex is for pleasure. In the best of all worlds, and from a public health perspective, it might be argued that the most important component of acceptability should be consideration of a method's efficacy in preventing unwanted outcomes. However, the case is that many women, including those at high risk, continue to have unprotected sexual experiences, resulting in STIs and unwanted pregnancies (Santelli et al., 2003). We believe that not all of these occasions involving the failure to protect can be blamed on the male partner, or pressure from the male partner. Simply put, often the behavioral episode is driven by a desire for pleasure with sensual enjoyment overriding considerations of prevention efficacy. Newcomer (2002) suggests, "understanding what makes prevention in the heat of the moment acceptable is essential to get a product used" (pg. 1). We believe that researchers and public health workers are more comfortable interacting with participants or patients around the issue of protection rather than sexual pleasure. Further, we suspect that the experience of pleasure, and what the interference of pleasure means, varies across cultures, relationships, locations, and studies. As argued elsewhere (Severy & Spieler, 2000), perhaps the challenge for new product development ought to be for innovators to begin with products that enhance sexual pleasure (there are many of these on the market), and then *add to the mix* features that prevent unwanted pregnancy and/or STIs.

Recent meta analyses of research investigating attitude-behavior models of condom use highlight the importance/salience of the impact of sexual enjoyment (Albarracin et al., 2001). Ellertson, Elias, and Coggins (1999) present findings highlighting the impact on sexual intimacy as a determinant of method use. In this case the method involved was early formulations of microbicides. Their work utilized both quantitative and qualitative procedures. Their results from three different cultures (Mexico, India, and Zimbabwe) generate similar conclusions. Namely, efficacy is not the sole determinant of acceptability and that when new methods are reviewed, attention to the impact on sexual intimacy is a requirement.

Recent and Forthcoming Methods

Although there has been a dearth of new methods in the last 25 or so years—perhaps due to the extreme cost of new method development—a number of innovations have been approved recently, or are about to be approved, for U.S. citizens. Our intent is to briefly identify seven of these approaches—and to review each of them with an eye to the acceptability issues discussed in this article.

Female Condom

Of the seven methods, the female condom is the most mature, having been approved by the Food and Drug Administration (FDA) in 1993 as a method to protect against STIs and unwanted pregnancies. In a recent review of research on the female condom, Gollub (2000) discusses four themes: (a) the female condom has empowered women in negotiating protection with male partners; (b) regulatory agencies and staff providers generate more constraints to use than does the woman's own reticence to master new knowledge and skills; (c) intervention trials must involve more support than simply handing the devices to women; and (d) the female condom is best suited to group or community level interventions. The greatest attention has been given to the first theme, the extent to which this method empowers women. Raphan, Cohen, and Boyer (2001) studied whether the female condom would be useful in empowering sexually active urban adolescent women. Their thought was that teens' high-risk behavior might be related to "personal factors of self-efficacy, sexual knowledge, self-esteem, and ability to communicate/negotiate" (p. 605). Their conclusion was that adolescent women will accept the female condom and can be empowered to protect themselves from STIs through its application or through the using of it as a negotiating tool. The Severy and Spieler (2000) argument is "that it is a lot easier for women to insert a female condom prior to a sex act . . . and then negotiate from the position of already having the condom in place, and keeping it inserted, in comparison to the negotiation of trying to persuade a man to put on a male condom" (p. 261). An interesting twist on this position is generated by the Rickert, Sanghvi, and Wiemann (2002) finding that adolescent women lack assertiveness. Adolescents appear to be a separate and distinct group of users—with their own culture, contexts, and issues worthy of study—including potential sexual violence when negotiation is unsuccessfully attempted.

The female condom is not immediately appealing to many people. Even the packaging material recommends at least three uses of the condom prior to making any judgements about how well liked it may be. An interesting study by Latka, Gollub, French, and Stein (in press) provides information regarding longer-term acceptability—the focus most relevant to this article. With U.S. STD clinic clients as participants, it was noted that the proportion of protected sex acts among women using the female condom at a six-month follow-up was nearly triple the baseline rate. It appears that acceptability of the female condom develops over time. Gollub (2000) concludes that the best liked features are: "(1) the fact that women can place it autonomously and can trust that it is not torn or otherwise sabotaged by the partner, (2) the high level of protection it can provide when used correctly, and (3) its soft and lifelike feel" (p. 1378). Alternatively, the least liked attributes include: the need to practice insertion, it can be seen by the partner, and the inner ring bothers some women (and their partners). Several of the earlier identified themes regarding acceptability are demonstrated by these data. Clearly, the nature of the relationship with the partner is crucial, and the impact on enjoyment is reflected in both positive and negative features. Artz, Demand, Pulley, Posner, and Macaluso (2002) demonstrated in a study of over 1100 women that the extent to which women have expressed their sexual likes and dislikes to their partners differentiates those women who have difficulty learning the proper insertion of the condom. Specifically, those feeling uncomfortable in expressing themselves were also those who had trouble with insertion.

Another aspect of the acceptability of the female condom involves the way in which women present the male partner with the prospect of use. According to Severy and Spieler (2000), some men report that if any condom were to be used, they would prefer to have sex with the woman wearing a female condom. It is not constricting to the man, is more "natural" for him, thereby making sex more pleasurable. Penman-Aguilar and associates (2002) found that men's initial response to the female condom did not predict continued use after the first trial. The authors argue that women must use multiple strategies, offering a variety of reasons for experimentation, to facilitate successful introduction of the female condom into a sexual relationship. Again, initial acceptability based on first exposure often is not related to long-term use.

Contraceptive Ring

Late in the fall of 2001, the FDA approved use of a vaginal ring (also known as the contraceptive ring), and it is now available to the public. The contraceptive ring is a hormonal method, with protection provided in a fashion not unlike oral contraceptives. The vaginal ring is a flexible, transparent device that is inserted into the vagina to prevent pregnancy for three weeks at a time, after which it is removed. When removed, it is discarded, and a new ring is inserted after the fifth day of the woman's menstrual cycle. The ring releases a continuous flow of estrogen and progestin into the woman's vagina, not into her system, and trials suggest that it is about 99% effective (*Network*, 2002). Clearly, the advantage over the pill is that one does not need to remember daily to take the pill, or carry the pills while traveling. Further, similar to the pill, it is under the woman's control and is not coitus dependent. One can discontinue the contraceptive action ring by simply removing it for more than three hours. An advantage over other methods such as Norplant is that there is no need to wait a year, or have surgically implanted hormone rods removed (Hardee, Balogh, & Villinski, 1997). This provides the woman (and her partner) with a much greater sense of control.

There are three types of rings. The first releases natural progesterone and is intended for breastfeeding women. The second type of ring uses a synthetic progestin, and is designed to be used for a year. It is a third type (estrogen and progestin) that has been approved by the FDA. It has fewer bleeding problems than the other rings, and offers cycle to cycle control. As is the case with oral contraceptives, vaginal rings provide no protection against STIs.

The contraceptive ring is two inches in diameter, and is about 1/8 inch thick. It can be easily removed, and re-inserted. The ring should not be taken out for more than three hours without backup protection. The removal feature carries with it interesting implications for intimate behavior. If the presence of the ring during the sex act bothers either the woman or her partner, it is easily removed on a temporary basis, and then reinserted after intercourse. Conversely, it is also possible that both men and women might enjoy the stimulating, extra contact between the penis and the ring, and the cervix and the "G spot" and the ring. This is a product that needs much more research as it reaches the Phase IV—the post-marketing stage—in its development.

Contraceptive Skin Patch

The FDA approved a contraceptive skin patch at almost the same time as the vaginal ring. It delivers a one-week supply of estrogen and progestin through the skin and systemically into the bloodstream, with many of the same qualities as the contraceptive ring. It allows the woman to be in control, not require a daily remembrance, allows for a change of fertility plans almost immediately, yet it does not provide protection from STIs.

There are two main differences between the ring and the patch. The former is active for three weeks (in comparison to one week) and the ring requires insertion into the vagina. However, the patch is worn continuously on the buttocks, lower abdomen, either the front or back of the upper torso, or the upper arms. The patch is removed each week and replaced with a new one for three weeks, and then the fourth week is "patch free" to allow for a regular menstrual period (*Network*, 2002). During development a variety of patch colors were tried. The first patch to be marketed is beige, and other formulations—including a clear patch—are being evaluated for further development (*Contraceptive Technology Update*, 2002).

The main behavioral acceptance implications involve the issue of adhesion, and observability. The woman wants the patch to stay on regardless of the activity—swimming, bathing, etc., but she needs to be able to remove the patch at the end of the week without problems or traces of where it has been. The other issue is much more related to the sexual experience. According to the clinical trial researchers (Nelson, 2002), some women actually "displayed it on the upper arm, kind of a 'badge of courage'" (pg. 3). In other words, the patch can be worn as a sign of protection from unwanted pregnancy. However, for those desiring privacy with their family planning behavior, or covert contraception with their partner, the patch may not be an option—since it must be worn to be effective.

Microbicides

In contradistinction to the previous two methods, the next two technologies are targeted predominantly to prevent unwanted sexually transmitted diseases—and if they happen to also protect against unwanted pregnancy, then they would have a dual action. The war against HIV/AIDS is being fought on many fronts. The largest biomedical attacks are under way in the form of attempts to develop microbicides and vaccines. Many feel that although condoms may be the best weapon against STIs, when their use cannot be negotiated or when they are not well liked, microbicides may be the best alternative (Feldblum, Weir, & Cates, 1999). Negotiation is simply not an option for many women in some cultures, particularly in developing countries where they do not have the social or economic status to demand condom use (Frank, Poindexter, Cox, & Bateman, 1995; Rosenberg & Gollub, 1992). This is not to suggest that negotiation should not, or will not, be involved with the use of microbicides. Rather, it is the case that, depending upon final versions of such a product, microbicides *may* offer the prospect of use without negotiation.

"Microbicides are agents that kill or deactivate disease-causing microorganisms, technically known as pathogens. Various types of pathogens cause STIs; therefore a microbicide that works against one pathogen does not necessarily work against others." (Boonstra, 2000, p. 3) Researchers are attempting to develop products that will be effective against a wide range of pathogens. Microbicide products could take the form of a film, gel, cream, or suppository. They would provide protection by blocking or killing the pathogens directly or preventing viral replication, should an infection occur. One of the more obvious target populations for the use of microbicides is discordant couples wherein one partner is seropositive for HIV and the other is not but they are in a stable union and wish to continue sexual relations.

Although there are almost 60 such products under development as of Spring, 2003, there have been no microbicides that have successfully survived the three phases of formal clinical trials necessary before commercial introduction. Most are in the early stages of testing (see FDA, n.d., *Home Page* or National Institute of Allergy and Infectious Diseases, n.d.). There are a number of Phase III clinical trials planned, and acceptability researchers are "waiting in the wings." As identified much earlier, the ground breaking "hypothetical acceptability" study in the United States by Darroch and Frost (1999) generated excitement as it indicated that there may be women interested in microbicides. Microbicide use can be initiated by the woman, she might attempt to use them clandestinely or covertly—avoiding negotiation, and they might not inhibit sexual activities. As for men, and to reflect a different culture, Ramjee, Gouws, Andrews, Myer, and Weber (2001) conducted interviews with South African men. They too, found a rather receptive audience for the hypothetical product. These researchers concluded that acceptability would depend upon the amount of lubrication and noninterference with sexual pleasure. In Zimbabwe (Moon, Khumalo-Sakutukwa, Heiman, Mbizvo, & Padian, 2002) yet other issues surfaced. Women were concerned about how microbicide use might affect their relationship with their husbands. Many men were concerned that women would be able to use the products without their consent or knowledge (in some situations this might be a good result for the public health of the larger community). In Zimbabwe, products will need to be introduced within the existing gender power structure. When all is said and done, hypothetical acceptability does not always translate to long-term use. Consequently, researchers have attempted to study the probable acceptability to microbicides by using proxies (products with many features similar to microbicides).

Elias and Coggins (1996) undertook an acceptability study of various spermicides as a proxy, and Family Health International conducted spermicide proxy studies in Mexico (Alvarado, Steiner, Spruyt, Joanis, & Glover, 1993), Kenya (Magiri et al., 1993), and the Dominican Republic (Cordero, Steiner, Spruyt, Joanis, & Glover, 1993) with the same strategy. Three different product formats were tested. Film products were preferred to gels and suppositories, which were thought to be too messy. Unfortunately, the films also generated resistance as they were thought to be too sticky. The challenge is that with vast cultural variation in acceptability, not to mention climate variations, microbicides may need to be provided in a variety of formats to be widely adapted. For example, in some African cultures "dry sex" is preferred, and in others it is not. Lubrication might carry only negative connotations in such an environment. Conversely, in western cultures such as the United States, there is a large market for products such as vaginal and penile lubricants. And, the use of lubricants in gay sexual behavior is extremely common.

Jones et al. (2001) report on a study in which women used barrier methods and N-9 spermicides. Acceptability was assessed at three and nine months post intervention. There were vast cultural differences, with Haitian women responding in a quite different fashion from other minority women (African American and Hispanic). Trial use of microbicides increased from 22% to 51% in three months, with no accompanying decrease in use of other barrier methods (e.g., male condoms) provided to the women.

Given the importance of microbicides as a potential weapon against STIs, much more research is needed on the acceptability of these products. Little or nothing is known about long-term use. What we do know seems to indicate that potential impact upon the couple relationship, and how this varies by culture, is going to be very important.

Vaccines

Another frontier for preventing HIV/AIDS involves vaccines. The concept is simple. Instead of a product that would be applied topically to a potential host site, can a vaccine be developed that would provide systemic protection? The National Institute of Allergy and Infectious Diseases is currently financing more than two dozen different possible vaccines in some stage of development (National Institute of Allergy and Infectious Diseases (n.d.)). In most cultures, people assume that vaccines are completely effective in preventing infections (e.g., Halloran, Longini, Haber, Struchiner, & Brunet, 1994). However, an HIV vaccine may turn out to be more like a flu vaccine—with efficacy rates less than 50%. The problem is that "the human immune system cannot turn back an HIV infection, and no one knows how to make a vaccine that accomplishes something the human body cannot do for itself" (Haney, 2001, p. 2). Perhaps new vaccines will moderate the disease without preventing it entirely. Consequently, scientists have been trying to lower the bar of expectations for the first generation of HIV vaccines, known as HIV Vaccine Trials Network (HVTN). The behavioral questions are, how will such vaccines be viewed by the public? what are the acceptability issues? and how might behavior (both individual and public health) be impacted by such products?

Consider the most positive features. A vaccine presumably would not interfere with any specific sexual act. The route of action is systemic, so protection would be available at all times and would allow for spontaneous behavior. There would be no observable sign of use (such as with the patch or a condom). This implies that protection could be clandestine—from both permanent and casual partners. This might be a method wherein some women could achieve protection in relationships wherein they typically have little, or no control at present.

Two important problems are easy to identify. First, vaccines would depend upon strong active agents. So, there is no way of knowing in advance the extent to which side effects might occur, at least for some people, under varying circumstances. Second is the rather universal belief (perhaps myth) that vaccines must, by their very nature, be extremely effective. If such beliefs are rampant, and if the first and/or second-generation HIV vaccines are more similar to flu shots than to those for childhood diseases like the mumps and measles, a catastrophe might ensue. The interaction of acceptability and vaccine efficacy again generates projected rates of protection, or lack thereof. Individuals believing that they are invulnerable may: (a) discontinue using other preventative methods—such as condoms, and (b) begin or resume much riskier sexual activities. Each of these may lead to even greater difficulties and rates of infection than before. An added note of complexity is that HIV mutates rapidly and becomes a moving target for any vaccine, even one that might be reasonably effective against one strain. Hence, any vaccine product might lead to greater public health problems if the vaccine is anything less than extremely effective. Paradoxically, vaccines may enjoy such high rates of acceptability, in spite of low efficacy, that their use generates health problems if multiple methods are not to be used.

Emergency Contraception

Technically speaking, emergency contraception (EC) is not a new method of contraception. However, this is an important method and its politicization has affected availability in the United States (Sherman, this issue). There are a number of acceptability issues generated by EC.

What about the market for EC? According to a USAID (1999) fact sheet, a wide variety of women at risk of pregnancy from unprotected intercourse can benefit from EC. These include women who have been raped, women whose partners' condoms break, women who run out of other contraceptive methods, women who forget to take several consecutive oral contraceptives, and other women who were not expecting to have sex. Is the intended market the same group as those willing to accept and use EC?

Two other questions are of interest. First, what happens to sexual behavior when EC becomes available? Second, what happens with regard to the use of other methods of protection when EC becomes available? Perhaps the knowledge that EC is available may result in higher rates of risky behaviors—especially as regards STIs. The impact knowledge has on the use of other methods is beginning to be understood. In separate studies in New York, Scotland, and Zambia, after EC use, the use of condoms as a routine method decreased (Glasier & Baird, 1998; "The impact," 1998). Another Zambian study (Skibiak, Ahmed, & Ketata, 1999) found that people stopped using other methods as they came to believe that EC could answer all of their needs. Another behavioral issue needs attention. Would advance provision of EC—having pills ahead of time, versus the more standard situation in which the pills are obtained after an *accident*, alter behavior, especially the use of the pills? The above Zambian study found that twice as many women correctly used the pills within 24 hours when they already had the pills in comparison with only having a prescription (remember that correct timing is crucial). Having the pills ahead of time may increase the very need for the pills. Alternatively, in the Scottish study, although one-time use increased with prior provision, women were not more likely to use EC repeatedly. Even more complexity in the issue involves whether or not to provide ECs in advance to men in men's clinics. Will this increase their availability to women, or decrease men's willingness to use condoms—especially if ECs are seen as a post-coital method that replaces condoms and interferes less with male pleasure? However, does it matter if ECs are used post-coitally—especially by adolescents and women who have infrequent sex—given the fact that there are no long-term effects? As we are interested in correct and consistent use as a definition of high acceptability, the situation with EC deserves much more research.

PERSONA

PERSONA is sold in England, Italy, France, Germany, France, and Ireland. Technically, it is not a contraceptive per se, rather it can thought of as a teaching device that informs the couple of the woman's fertile period. The system is designed to allow the couple to make decisions regarding unprotected sex by having knowledge of the woman's fertility status that day. As with many of the methods discussed here, PERSONA does not offer any protection against STIs. In essence, the device can be thought of as an extension and enhancement of natural family planning methods. The developers, Unipath, had conducted market research (in essence, hypothetical acceptability research) with over 3000 women from a number of countries regarding a concept that depends upon personal hormone monitoring as a contraceptive method. PERSONA depends upon the monitoring of estrone-3-glucuronide (E3G) and luteinizing hormone (LH), which are the best markers to determine the fertile period. Utilizing frequently updated information, in addition to six months of stored personal data, PERSONA identifies the days of each cycle a woman can have intercourse without risk of pregnancy (or, conversely, days on which she should protect herself with other methods). The efficacy of PERSONA has been reported to be 94% by Bonnar, Flynn, Freundl, Kirkman, Royston, and Snowden (1999) and the cost-effectiveness has been demonstrated by Gambone and Tabbush (2002).

Clearly, the concept of alternatives is important here. We have argued that true acceptability implies choice, not forced use. For those not at risk of STIs, with a history of using barrier methods, PERSONA allows the freedom to make such behavior unnecessary on all but the fertile days of the cycle. For couples with a history of systemic protection, knowing that there are days on which no method is required allows freedom from the systemic. Alternatively, knowing that additional protection is required on other days might seem burdensome. Acceptability often depends upon the couple's knowledge of alternatives and experience with them. At the end of six months of observation with 400 English couples who were using the monitor, more than 50% reported that their sex life had gotten better, another 40% or so stayed the same, while about 7% reported that it had gotten worse. The impact a method has on one's perception of one's sex life is critically related to the overall acceptability of a method (Severy & Spieler, 2000).

Another study was conducted by the first author (Severy, McNulty, Findley-Klein, & Robinson, 2004) which compared the acceptability of condoms and PERSONA with couples from California as part of a Phase III randomized clinical trial. In this case, couples interested in the study were randomly assigned into one of two conditions—use PERSONA for one year, or use condoms for one year. Couples completed questionnaires after the 1st, 3rd, 4th, 6th, and 13th menstrual cycles. Data were collected regarding overall levels of acceptability, the product's attributes, the perceptions that one's partner liked the method, the impact on partner communication regarding sexual and fertility decision making, the quality of their overall relationship, one's self esteem, and the impact on the couple's intimate behavior. Data were collected from both the male and the female partner.

Couples using PERSONA were much more likely to report positive sexual experiences with their partners than were those assigned to the condom use arm of the study. Further, there is a small, but discernable, growth in ratings over time for the PERSONA couples. Clearly, if long-term and consistent use is the criterion for acceptability of a method, this device has attracted advocates for its availability in the United States. Note, however, that both of these trials included couples in relatively stable relationships only. This may be a rather small niche, and given that PERSONA does not protect against STIs, its overall attractiveness may be limited.

Concluding Reflections

We have tried to present the case that acceptability is a complex issue, and that the response (including use) to any method designed to prevent STIs and/or unwanted pregnancies is multi-faceted. It seems to us that the most meaningful way to consider acceptability involves voluntary, consistent, correct, and long-term use. Confidence in indices of a method's acceptability increase if there are available alternatives that women and their partners choose NOT to use. Further, we believe confidence in ratings of acceptability increase even further if method use continues across situations and cultures. Perhaps the most dramatic evidence of acceptability would be consistent and correct use during *the heat of the moment*. Although by studying the behavioral response and implications of new methods it is often possible to *infer* acceptability, direct assessment remains a challenge.

References

Albarracin, D., Ho, R., McNatt, P., Williams, W., Rhodes, F., Malotte, C. K., et al. (2001). The structure of outcome beliefs in condom use. *Health Psychology, 19*, 458–468.

Alvarado, G., Steiner, M., Spruyt, A., Joanis, C., & Glover, L. (1993). *Contraceptive film acceptability study: Mexico*. Research Triangle Park, NC: Family Health International.

Amaro, H. (1995). Love, sex and power: Considering women's realities in HIV prevention. *American Psychologist, 50*, 437–447.

Artz, L., Demand, M., Pulley, L. V., Posner, S. F., & Macaluso, M. (2002). Predictors of difficulty inserting the female condom. *Contraception, 65*, 151–157.

Becker, S. (1996). Couples and reproductive health: A review of couple studies. *Studies in Family Planning, 27*, 291–306.

Bonnar, J., Flynn, A., Freundl, G., Kirkman, R., Royston, R., & Snowden, R. (1999). Personal hormone monitoring for contraception. *The British Journal of Family Planning, 24*, 128–134.

Boonstra, H. (2000, February). Campaign to accelerate microbicide development for STD prevention gets under way. *The Guttmacher Report on Public Policy, 3*(1), 3–4, 14.

Brady, M. (2003). Preventing sexually transmitted infections and unintended pregnancy, and safe-guarding fertility: Triple protection needs of young women. *Reproductive Health Matters, 11*, 134–141.

Burns, L. H. (2003). *Cross-cultural issues and the use of assisted reproductive technologies*. Unpublished paper. University of Minnesota Medical School, Minneapolis, MN.

Castle, S., Konate, M. K., Ulin, P. R., & Martin, S. (1999). Clandestine contraceptive use in urban Mali. *Studies in Family Planning, 30*, 231–248.

Catania, J., Kegeles, S., & Coates, T. (1990). Towards an understanding of risk behavior: An AIDS risk reduction model (ARRM). *Health Education Quarterly, 17*, 53–72.

Cleland, J., Hardy, E., & Taucher, E. (1990). *Introduction of new contraceptives into family planning programmes: Guidelines for social science research*. Geneva, Switzerland: World Health Organization.

Contraceptive ring, patch approved. (2002). *Network 21*(2), 27.

Cordero, M., Steiner, M., Spruyt, A., Joanis, C., & Glover, L. (1993). *Contraceptive film acceptability study: Dominican Republic*. Research Triangle Park, NC: Family Health International.

Darroch, J. E., & Frost, J. J. (1999). Women's interest in vaginal microbicides. *Family Planning Perspectives, 31*, 16–23.

Davidson, A. R., & Jaccard, J. J. (1979). Variables that moderate the attitude behavior relation: Results of a longitudinal survey. *Journal of Personality and Social Psychology, 37*, 364–376.

Davidson, A. R., & Morrison, D. M. (1983). Predicting contraceptive behavior from attitudes: A comparison of within- versus across-subjects procedures. *Journal of Personality and Social Psychology, 45*, 997–1009.

Elias, C., & Coggins, C. (1996). Female-controlled methods to prevent sexual transmission of HIV. *AIDS, 10*, S43–S51.

Elias, C., & Coggins, C. (2001). Acceptability research on female-controlled barrier methods to prevent heterosexual transmission of HIV: Where have we been? Where are we going? *Journal of Women's Health & Gender-Based Medicine, 10*, 163–173.

Ellertson, C., Elias, C., & Coggins, C. (1999, October). *Current research on microbicides acceptability*. Paper presented at the National Institutes of Health, Bethesda, MD.

Feldblum, P. P., Weir, S. S., & Cates, W., Jr. (1999). The protective effect of condoms and nonoxynol-9 against HIV infection: A response to Wittkowski and colleagues. *American Journal of Public Health, 89*, 108–110.

Food and Drug Administration. (n.d.). *Home Page*. Retrieved June 15, 2002, from http://FDA.GOV.

Food and Drug Administration. (n.d.). *Device Evaluation Information*. Retrieved March 30, 2003, from http://www.FDA.GOV/CDRH/ODE/WHATWEDO.HTML.

Frank, M. L., Poindexter, A. N., Cox, C. A., & Bateman, L. (1995). A cross-sectional survey of condom use in conjunction with other contraceptive methods. *Women's Health Journal, 23*, 31–46.

Gambone, J., & Tabbush, V. (2002, April 16). *Cost-effectiveness of a fertility monitoring device*. Presented at "Healthcare Financing for the 21st Century," University of Maryland.

Glasier, A., & Baird, D. (1998). The effects of self-administering emergency contraception. *New England Journal of Medicine, 339*, 1–4.

Gollub, E. L. (2000). The female condom: Tool for women's empowerment. *American Journal of Public Health, 90*, 1377–1381.

Good, B. J. (1977). The heart of what's the matter: The semantics of illness in Iran. *Culture, Medicine and Psychiatry, 1*, 25–58.

Halloran, M. E., Longini, I. M. Jr., Haber, M. J., Struchiner, C. J., & Brunet, R. C. (1994). Exposure efficacy and change in contact rates in evaluating prophylactic HIV vaccines in the field. *Statistics in Medicine, 13*, 357–377.

Haney, D. Q. (2001, August 25). *AIDS turnaround: Researchers suddenly upbeat over vaccine prospects.* Retrieved August 27, 2001, from the Associated Press Web site: http://www.ap.org

Hardee, K., Balogh, S., & Villinski, M. T. (1997). Three countries' experience with Norplant introduction. *Health Policy and Planning, 12*, 199–213.

Hardy, E., de Padua, K. S., Jimenez, A. L., & Zaneveld, L. J. (1998). Women's preferences for vaginal antimicrobial contraceptives: Preferred characteristics according to women's age and socioeconomic status. *Contraception, 58*, 239–244.

HIV Vaccine Trials Network. Retrieved February 20, 2003, from http://www.HVTN.ORG/Science/trials.html.

The impact of patient experience or practice: The acceptability of emergency contraceptive pills in inner-city clinics. (1998). *Journal of the American Medical Association, 53*, 255–257.

Jones, D. L., Weiss, S. M., Malow, M. I., Devieux, J., Stanley, H., Cassells, A., et al. (2001). A brief sexual barrier intervention for women living with AIDS: Acceptability, use, and ethnicity. *Journal of Urban Health, 78*, 593–604.

Keller, A. (1979). Contraceptive acceptability research: Utility and limitations. *Studies in Family Planning, 10*, 230–241.

Latka, M., Gollub, E. L., French, P., & Stein, Z. (in press). Male and female condom use among women after counseling in a risk reduction hierarchy for STD prevention. *Sexually Transmitted Diseases.*

Magiri, B., Onoka, C., Trangsrud, R., Steiner, M., Spruyt, A., Joanis, C., et al. (1993). *Contraceptive film acceptability study: Kenya.* Research Triangle Park, NC: Family Health International.

Marshall, J. (1977). Acceptability of fertility regulating methods: Designing technology to fit people. *Preventive Medicine, 6*, 65–73.

Miller, W. B., Severy, L. J., & Pasta, D. J. (2004). A framework for modeling fertility motivation in couples. *Population Studies, 58*, 193–205.

Minnis, A. M., Shiboski, S. C., & Padian, N. S. (2003). Barrier contraceptive method acceptability and choice are not reliable indicators of use. *Sexually Transmitted Diseases, 30*, 556–561.

Moon, M. W., Khumalo-Sakutukwa, G. N., Heiman, J. E., Mbizvo, M. T., & Padian, N. (2002). Vaginal microbicides for HIV/STD prevention in Zimbabwe: What key informants say. *Journal of Transcultural Nursing, 13*, 19–23.

National Institute of Allergy and Infectious Diseases. (n.d.) *Home Page.* Retrieved April 23, 2003, from http://www.NIAID.NIH.GOV

Nelson, A. (2002). First contraceptive patch offers once-a-week dosing. *Contraceptive Technology Update Vol. 23*, 1–3.

Newcomer, S. (2002, May 12–15). *Rapporteur: Acceptability studies.* Paper presented at Microbicides 2002 Conference, Antwerp, Belgium.

Padian, N. (1999, October). *Measurement issues.* Paper presented at the National Institutes of Health, Bethesda, MD.

Painter, T. M. (2001). Voluntary counseling and testing for couples: A high-leverage intervention for HIV/AIDS prevention in sub-Saharan Africa. *Social Science & Medicine, 53*, 1397–1411.

Penman-Aguilar, A., Hall, J., Artz, L., Crawford, M. A., Peacock, N., van, Olphen, J., Parker, L., & Macaluso, M. (2002). Presenting the female condom to men: A dyadic analysis of effect of the woman's approach. *Women and Health, 35*, 37–51.

Ramjee, G., Gouws, E., Andrews, A., Myer, L., & Weber, A. E. (2001). The acceptability of a vaginal microbicide among South African men. *International Family Planning Perspectives, 27*, 164–170.

Raphan, G., Cohen, S., & Boyer, A. M. (2001). The female condom, a tool for empowering sexually active urban adolescent women. *Journal of Urban Health, 78*, 605–613.

Rickert, V. I., Sanghvi, R., & Wiemann, C. M. (2002). Is lack of assertiveness among adolescent and young women a cause for concern? *Perspectives on Sexual and Reproductive Health, 34*, 178–183.

Rosenberg, M., & Gollub, E. (1992). Commentary: methods women can use that may prevent sexually transmitted disease, including HIV. *American Journal of Public Health, 82*, 1473–1478.

Santelli, J., Rochat, R., Hatfield-Timajchy, K., Gilbert, B. C., Curtis, K., Cabral, R. et al. (2003). The measurement and meaning of unintended pregnancy. *Perspectives on Sexual and Reproductive Health, 35*, 94–101.

Severy, L. J. (1999). Acceptability as a critical component of clinical trials. In L. J. Severy & W. B. Miller (Eds.), *Advances in population: Psychosocial perspectives, Vol. 3* (pp. 103–122). London: Jessica Kingsley Publishers.

Severy, L. J., McNulty, J., Findley-Klein, F., & Robinson, J. (2004). *Assessing the acceptability of innovative contraceptive technology: Participation profiles and implications for clinical trial success.* Manuscript submitted for publication.

Severy, L. J., & Silver, S. E. (1993). Two reasonable people: Joint decision-making in fefrtility regulation. In L. J. Severy (Ed.), *Advances in population: Psychosocial perspectives, Vol. 1* (pp. 207–227). London: Jessica Kingsley Publishers.

Severy, L. J., & Spieler, J. (2000). New methods of family planning: Implications for intimate behavior. *Journal of Sex Research, 37*, 258–265.

Severy, L. J., & Thapa, S. (1994). Preferences and tolerance as determinants of contraceptive acceptability. In L. J. Severy (Ed.), *Advances in population: Psychosocial perspectives, Vol. 2* (pp. 207–227). London: Jessica Kingsley Publishers.

Skibiak, J. P., Ahmed, Y., & Ketata, M. (1999). *Testing strategies to improve access to emergency contraception pills: Prescription vs prophylactic distribution.* Nairobi, Kenya: The Population Council, Africa OR/TA Project.

Snowden, R. (1996). Human reproductive behavior: General principles. In J. Bonnar (Ed.), *Natural contraception through personal hormone monitoring: Proceedings of a symposium held at the XIV Figo World Congress of Gynecology and Obstetrics* (pp. 45–53). New York: Parthenon Publishing Group.

Spieler, J. (1997, March). *Behavior issues in dual protection.* Paper presented at the Psychosocial Workshop, Population Association of America, Washington, DC.

Stein, Z. (1996). Family planning, sexually transmitted diseases, and the prevention of AIDS—divided we fail? *American Journal of Public Health, 86*, 783–784.

Thomson, E., & Hoem, J. (1998). Couple childbearing plans and births in Sweden. *Demography, 35*, 315–322.

Tolley, E. L., Loza, S., Kafafi, L., & Cummings, S. (2003). *The impact of menstrual changes on method use: A comparative study among women using the IUD, DMPA and Norplant.* Research Triangle Park, NC: Family Health International.

Treiman, K., Liskin, L., Kols, A., & Rinehart, W. (1995). IUDs—An Update. *Population Reports, 23*(5) PP.

UNAIDS. (2002). *Report on the global HIV/AIDS epidemic.* The Barcelona report. New York: Author.

USAID. (1999). *USAID fact sheet on emergency contraception.* Unpublished document provided by J. Spieter, Chief, Research Division, office of Population, United States Agency for International Development. Washington, DC.

Westoff, C. F., & Ryder, N. B. (1977). The predictive validity of reproductive intentions. *Demography, 14*, 431–453.

What do women want? The answer is simple: Convenient birth control. (2002). *Contraceptive Technology Update,* (23), 13–15.

LAWRENCE (LARRY) J. SEVERY obtained his PhD in social psychology at the University of Colorado in 1970. He spent the summer of 1976 at the University of North Carolina participating in one of National Institute of Child Health and Human Development's (NICHD) post-doctoral training programs for population research. His long-term interests have focused upon couple decision-making regarding family planning and contraception, the acceptability of new technologies and programming aimed at preventing unwanted pregnancies or sexually transmitted diseases—especially HIV, and program evaluation. Larry is the R. David Thomas Professor of Psychology at the University of Florida, and served as the 2002–3 President of Division 34—Population and Environmental Psychology—of the American Psychological Association. Currently, he is the Director of Behavioral and Social Sciences, Family Health International (FHI), Research Triangle Park, NC, by virtue of a long-term contract between FHI and the University of Florida.

SUSAN F. NEWCOMER is a statistician/demographer at National Institutes of Health (NIH) in the Demographic and Behavioral Sciences Branch of NICHD. She holds a 1983 PhD in sociology and population studies from the University of North Carolina, and a 1962 BA from Barnard in psychology and Chinese. From 1984–1988, she was the Director of Education, Planned Parenthood Federation of America. Before obtaining her degree she had been acting director of a YWCA on a college campus, and director of a daycare center. At NIH she is the project officer on a portfolio of research grants on fertility, HIV risk, contraceptive use, acceptability research, reproductive health and adolescent risk behaviors, including interventions to reduce risk in youth and adults. Another of her responsibilities is to develop and maintain contacts with behavioral and social science researchers in the United States and in other countries, and to provide assistance to those wishing to apply for funding to the NIH. Her publications have dealt with adolescent sexuality, fertility, and sexuality education.

Journal of Social Issues, Vol. 61, No. 1, 2005, pp. 67–93

Context of Acceptability of Topical Microbicides: Sexual Relationships

Helen P. Koo[*]

RTI International

Cynthia Woodsong

Family Health International

Barbara T. Dalberth

RTI International

Meera Viswanathan

RTI International

Ashley Simons-Rudolph

RTI International

Domains central to the effects of sexual relationships on the acceptability of a vaginal protection method were explored in 14 focus groups and 38 in-depth interviews with women and men recruited from a health department's sexually transmitted infections (STI) and family planning clinics. Findings indicate that acceptability depended on a couple's relationship type, classified as serious, casual, or "new." Potential barriers to communication about product use may be overcome through direct or indirect covert use, depending on relationship type. More men than women thought women should always tell their partners if they use microbicides, regardless of relationship type. Results indicate the importance of the relationship context

[*]Correspondence concerning this article should be addressed to Helen P. Koo, RTI International, 3040 Cornwallis Road, Research Triangle Park, NC, 27709 [e-mail: hpk@rti.org].

The research reported in this article was funded by grant RO1 – HD040141 from the National Institute of Child Health and Human Development. The authors gratefully acknowledge the roles of Peter Leone, M.D., as co-investigator, and Deb Norton, M.D., Gibbie Harris, Peter Morris, M.D., Ida Dawson, and Karen Best in facilitating data collection at Wake County Human Services, and the work of the staff who conducted the focus groups and interviews.

in understanding the likely acceptability of using microbicides, and perhaps any method of STI/HIV protection.

The incidence of Human Immunodeficiency Virus (HIV) infection is growing faster for women than men, and the major cause of infection for women is heterosexual contact. Women now comprise half of HIV infections worldwide (Annan, 2004; Population Council, 2001); they make up the fastest growing population of new HIV infections (Malow, 2001). In the United States, Acquired Immune Deficiency Syndrome (AIDS) is the fourth leading cause of death for women ages 25 to 44 (National Institute of Allergy and Infectious Disease [NIAID], 2000), and the most frequent source of exposure is heterosexual contact across all race and ethnic groups (Centers for Disease Control and Prevention [CDC], 2003). Although condoms could protect women from HIV infection, many women cannot ensure that their male partners use the male condom nor gain their consent to use female condoms (e.g., Beadnell, Baker, Morrison, & Knox, 2000; Karim, Karim, Soldan, & Zondi, 1995; Lauby, Semaan, O'Connell, Person, & Vogel, 2001; Luke, 2003; Mill & Anarfi, 2002; Pulerwitz, Amaro, De Jong, Gortmaker, & Rudd, 2002; Tang, Wong, & Lee, 2001). Apart from an HIV vaccine, hope for the protection of many women rests on developing topical or vaginal microbicides.

Microbicides are chemical substances that, put into a carrier such as a gel, suppository, cream, or film, could be inserted into the vagina or rectum to prevent or reduce transmission of sexually transmitted infections (Alliance for Microbicide Development, 2003). Various forms being tested would work by destroying the pathogen, blocking its entry into or fusion with target cells, or inhibiting its replication once inside the target cell (Rockefeller Foundation, n.d.-b). While more than 50 products are in development, 16 are in clinical trials, including approximately 8 have been tested in Phase 1 (pharmacokinetic and safety); 4 in Phase 2 (efficacy); and 4 are expected to be in Phase 3 (expanded safety and efficacy) trials in the near future (Alliance for Microbicide Development, 2004). This is an opportune time to investigate the acceptability of vaginal microbicides to potential users: Findings could be obtained in time to help formulate the product characteristics and develop different ways to present or market the product to appeal to various high risk groups.

Topical microbicides have several features that confer significant advantages over condoms: they allow (a) skin-to-skin contact during sexual intercourse and (b) the possibility of discrete or even covert use without the partner's knowledge, consent, or cooperation, as well as (c) use with male condoms for additional protection. Because of the special needs of women who cannot obtain their partners' cooperation in using condoms, the possibility of covert use is one of the most promising features of microbicides. This article reports on findings on the role of relationship factors in men and women's perceptions of vaginal microbicides derived from the qualitative phase of a study conducted in North Carolina.

Importance of Relationship Factors

Existing theories of behavior change that seek to explain the adoption and practice of STI/HIV protection focus on individuals (e.g., Transtheoretical Model [Prochaska, Redding, Harlow, Rossi, & Velicer, 1994] and the AIDS Risk Reduction Model [Catania, Kegeles, & Coates, 1990]); they do not take into account broader issues driving risk behavior (e.g., Fisher & Fisher, 2000; Harvey, 2001; Logan, Cole, & Leukefeld, 2002; Miller & Neaigus, 2001). Chief among these issues is that sexual relationships involve two people who intimately interact with each other. The nature of these relationships may influence women's perception of the need for protection, as well as their willingness and ability to protect themselves. Even if a woman recognizes that she is at risk of infection by her partner, in some cases she may not be able to act independently of him to protect herself.

Researchers have recently called for and conducted empirical studies concerning the effects on women's HIV risk of gender roles, relationship dynamics, and various dimensions of relationship power and their basis in broader structural and cultural forces (Amaro, 1995; Blanc, 2001; Castañeda, 2000; Harvey, Beckman, Browner, & Sherman, 2002; Harvey, Bird, Galavotti, Duncan, & Greenberg, 2002; Impett & Peplau, 2003; Jenkins, 2000; Jewkes, Levin, & Penn-Kekana, 2003; Miller & Neaigus, 2001; O'Leary, 2000; Parker, 2001; Parker, Easton, & Klein, 2000; Pulerwitz, Gortmaker, & DeJong, 2000; Pulerwitz et al., 2002; Quina, Harlow, Morokoff, & Burkholder, 2000; Reid, 2000; Sherman, Gielen, & McDonnell, 2000; Simoni, Walters, & Nero, 2000; Vanwesenbeeck, van Zessen, Ingham, Jaramazovic, & Stevens, 1999; Wingood & Diclemente, 2000). As these authors point out, relationship issues such as nature or type of relationship, relative power, trust, commitment, emotional closeness, satisfaction with the relationship, the importance of sexual intercourse, communication about sexual history and behavior, decision-making about sexual matters and protection, and partner abuse influence all aspects of the decision to use a product (as well as subsequent use adherence).

Research on condom acceptability also suggests the importance of these factors (DiClemente, 1991; Ford & Norris, 1995; Hingson, Strunin, & Berlin, 1990; Hingson, Strunin, Berlin, & Heeren, 1990; Nathanson, Upchurch, Weisman, Kim, Gehret, & Hook, 1993; Overby & Kegeles, 1994). In particular, one of the most consistent findings in the literature on condoms is that they are more likely to be used with casual partners than with the main or steady partner, and in the beginning of relationships (e.g., Anderson, Wilson, Doll, Jones, & Barker, 1999; Carroll, 1991; Ellen, Cahn, Eyre, & Boyer, 1996; Ford & Norris, 1993; Macaluso, Demand, Artz, Fleenor, Robey, Kelaghan, & Cabral, Hook, 2000; Plichta, Weisman, Nathanson, Ensminger, & Robinson, 1992; Sacco, Rickman, Thompson, Levine, & Reed, 1993; Santelli, Kouzis, Hoover, Polacsek, Burwell, & Celentano, 1996; Woodsong & Koo, 1999). Furthermore, concerns about partner reactions and communication about method use have been found to influence the acceptability of female condoms (Bogart, Cecil, & Pinkerton, 2000; Choi, Gregorich, Anderson, Grinstead, & Gomez, 2003; Hirky, Kirshenbaum, Melendez, Rollet, Perkins, & Smith, 2003), which like topical microbicides, are applied by the woman. Recently, authors have called attention to other female methods, such as the diaphragm and cervical cap, as possibly offering protection against STIs/HIV (Ellertson & Burns, 2003; Harvey, Bird, & Branch, 2003). They point out that men's attitudes and reactions to the use of these methods could affect their acceptability to women as well (Harvey et al., 2003; Powell, Mears, Deber, & Ferguson, 1986).

Many authors have noted that gender roles, which are culturally prescribed and sanctioned, shape women's views of the centrality of their relationships to their identities and may place them in positions of unequal power vis-à-vis their male partners (e.g., Bowleg, Belgrave, & Reise, 2000; Christianson, Johansson, Emmelin, & Westman, 2003; Gutierrez, Oh, & Gillmore, 2000; Pulerwitz et al., 2000; Pulerwitz et al., 2002; Reid, 2000; Seage, Holte, Gross, Koblin, Marmor, Mayer, & Lenderking, 2002; Sherman et al., 2000; and Simoni et al., 2000). Evidence suggests that the higher levels of risk behavior observed among African American and Hispanic men than Caucasian men (Choi, Catania, & Dolcini, 1994) may be supported by social and cultural norms (Logan et al., 2002). Norms which tolerate or even accept male high-risk behavior among these racial and ethnic groups, coupled with some women's economic dependence on partners (particularly among low-income women) and susceptibility to intimate partner violence make issues of control and empowerment central to the acceptability of a new protection product.

Study Design

Our study, "Acceptability of Microbicides across Risk Groups and Time," investigates a wide range of product attributes and individual, relationship, and sociocultural factors that are likely to affect the adoption and continued use of microbicides. To investigate possible differences across social and cultural groups, participants included African Americans, Latinos (nearly all of whom were recent immigrants from Mexico), and Caucasians. The study includes two phases. The first phase used primarily qualitative methods, aimed at exploring domains identified primarily in the literature and developing survey instruments that include more replicable measures of acceptability and associated factors. In the second phase, we implemented a short-term longitudinal survey using these instruments.

The present article reports results based on data collected during the qualitative phase. Data were collected from mid-2001 to mid-2002, primarily from female clients seen in the STI and family planning clinics of a large public health department in North Carolina, their male partners, and other men.

We used multiple qualitative methods to obtain a deeper understanding of women's and men's perceptions and experiences that relate to acceptability of vaginal microbicides. To develop familiarity with the social and cultural environment of the research subjects, we first conducted interviews with 23 health professionals to obtain their understanding of their clients' STI/HIV prevention attitudes, behaviors, and likely attitudes toward vaginal microbicides. We also conducted participant observations of the clinics to understand clinician-patient interactions. We used the insights obtained from the clinicians and observations to help us develop a focus group guide to supplement and elaborate on topics suggested by the literature on HIV prevention.

We conducted focus groups to explore domains of meaning and community norms and values (Krueger, 1994; Morgan, 1993) that influence "safe sex" behaviors and use of products inserted in the vagina. Since the focus group method is not suited to collecting data on personal experience of a sensitive nature (Seidman, 1991), we also conducted a number of individual, in-depth interviews with people who had not participated in the focus groups to collect personal data on many of the same issues. We used focus group data to develop the interview guides.

At the end of each focus group and in-depth interview, we collected quantitative data from participants in a short, self-administered questionnaire.

Topics of Inquiry

The focus group and interview guides included a wide range of topics, a subset of which are reported in the present article; these are listed in Table 1. Based on the literature reviewed above, we expected that many of the responses to our questions about future microbicide products would be contingent on the type and nature of relationships in which they would be used. Thus, we explored types of sexual relationships and norms for communication within those relationships, followed by more specific discussions about vaginal microbicides (explained as a future method that women could use) within the context of relationships.

The short survey asked questions about sociodemographic characteristics, perception of STI/HIV risk, and if women should tell their partners if they use vaginal protection products.

Table 1. Topics of Inquiry

Relationship types and related perceptions about STI/HIV transmission risks
Relationship types and communication about contraceptive use
Relationship types and communication about and need for STI/HIV prevention
Relationship types and likely reaction to the woman's request to use a vaginal microbicide
General reactions to the concept of covert use of vaginal microbicides
Views on whether people should and would talk about use of a vaginal microbicide
Views on how and when a woman should inform her partner about microbicide use
Reasons why women would not want to tell a partner about microbicide use
Perceived male reaction to woman's decision to use a microbicide
Perceived male reaction to discovery of covert microbicide use
Perceived responsibility for microbicide use

Table 2. Type and Number of Data Collection Units

	Female Groups or Individuals[1]			Male Groups or Individuals[1]		
Data Collection Type	African American	Caucasian	Latina	African American	Caucasian	Latino
Focus Groups						
Adults	2	2	2	2	1	1
Teens	1	1	0	1	1	0
Interviews						
Adults	7	3	6	5	4	3
Teens	4	3	0	3	0	0
Short Survey						
Adults	23	11	19	24	13	15
Teens	9	7	0	7	5	0

[1]For focus groups, the numbers refer to the number of groups; for interviews and short survey, the numbers refer to the number of individuals.

Recruitment

We recruited adult women and men and adolescents of both genders (ages 16–19). To be eligible to participate, females and males had to have engaged in heterosexual intercourse on average at least once a week. Recruitment of both men and women was conducted at STI and family planning clinics at the local urban health department. Potential participants were either approached directly by study staff, referred by clinic staff or another study participant (including female partners), or they responded to flyers posted in the health department. Participants were told that they would be reimbursed $50 for their time. Table 2 shows the numbers of focus groups (14), and participants in interviews (38) and the short survey (133), by gender and race/ethnicity. Of the 15 males interviewed, 8 were partners of the women interviewed.

Data Collection

To ground the discussion of microbicides in actual use experience, we gave women two forms of vaginal lubricants that use the same delivery systems as those likely to be used for microbicides—a gel (Replens) and a suppository (Lubrin)—and asked them to use each at least once during sexual intercourse before participating in the focus groups or interviews. Both lubricants are over-the-counter products that offer no protection against STIs, HIV, or pregnancy.

To facilitate discussion of sensitive topics, focus groups were homogeneous in race/ethnicity, gender, and age (adult or adolescent). Focus group moderators and in-depth interviewers were matched by sex with the participants, and, when possible, by race. We slightly modified focus group and interview guides to be sensitive to gender and age differences, and we conducted data collection with Latino participants in Spanish. Latino participants were given the choice of the language they preferred. A few chose English and were put with the English-speaking Caucasian focus groups or were interviewed in English. The health department has recently experienced a great increase in the number of Latino clients, nearly all recent immigrants, who speak little English.

Staff experienced in qualitative data collection conducted the focus groups and interviews. The study's senior qualitative research staff member conducted two training workshops for all data collection staff—one on conducting focus groups and one on in-depth interviews—using the instruments designed for the study and methods standards adopted for the study. In the focus group training, all staff practiced moderating a mock focus group discussion with other staff members. In the interview training, all staff were required to conduct a full tape-recorded interview with another member of the parent research institution (RTI), or a personal friend or acquaintance. The senior qualitative staff member then reviewed these interviews for quality, and her observations were reviewed at a final team meeting before data collection began.

We held focus groups in the health department conference rooms and conducted in-depth interviews primarily in the women's homes. If we also interviewed their male partners, female and male interviewers either conducted the interviews concurrently in separate locations of the home, or the male interviews were scheduled separately at a location convenient for the male participant.

We followed standardized procedures accepted for qualitative data to capture and analyze the data (Bernard, 2000; Denzin & Lincoln, 2000; Miles & Huberman, 2000; Ulin, Robinson, Tolley, & McNeill, 2004). We audiotaped focus group discussions, transcribed them, and checked transcriptions for accuracy. Spanish transcripts were translated into English and reviewed by the moderator to ensure accurate capture of data. We also audiotaped the individual interviews but did not transcribe them. Instead, the interviewer took detailed notes during the interviews and later expanded his or her notes by listening to the audiotapes. Staff who conducted Spanish interviews translated their notes into English and wrote the interview report in English.

All study procedures, including the informed consent forms and process, were approved by the Institutional Review Boards (IRBs) of RTI and the University of North Carolina at Chapel Hill. (The health department had an agreement with the University to use the latter's IRBs.) Women were administered a written informed consent, during which they had to be able to say in their own words that (a) the vaginal lubricants offered no protection against STIs, HIV, or pregnancy, (b) they understood the need to use the condoms provided to protect against STIs/HIV, and (c) they should continue to use their usual birth control methods if they did not wish to become pregnant. They then initialed written statements to this effect and signed the consent form. Because they were not given any products to use, men were administered a less elaborate, verbal informed consent.

Data Analysis

We used a grounded theory approach (Glaser & Strauss, 1967; Strauss & Corbin, 1990), with inductive and deductive coding to identify and explore themes that were identified in an iterative fashion throughout the initial qualitative research period (Bernard, 2000; Miles & Huberman, 2000; Spradley, 1979). This approach allowed us to examine responses to direct queries as well as emergent concepts and response categories. The coded data from multiple types of data sources (e.g., focus groups, in-depth interviews) could then be examined together, to provide additional depth of meaning. Our article reports findings derived from this analysis approach. Since not all participants responded to questions about the emergent issues, it is inappropriate to attempt to quantify the qualitative results reported here. Rather, throughout the article we use "exemplar quotes" (Bernard, 2000, p. 444), to better illustrate aspects of a particular finding and the perspectives of our participants.

We imported all focus group transcripts and interview notes into QSR NVIVO software (Richards, 2002), a qualitative analysis software package used for coding and analysis.

A team of five analysts coded all documents. An experienced qualitative analyst served as coding supervisor and was responsible for training and overseeing the work of four additional analysts. As noted above, the team conducted the qualitative coding using a mixed deductively and inductively-oriented approach to the application of codes to lines of text data. The coding enabled efficient grouping and comparison of data. The full analysis study team participated in the development of a project codebook (MacQueen, McLellan, Kay, & Milstein, 1998), which reflected both a priori codes created in response to specific research questions and additional codes created in response to themes that emerged from the data. The codebook provided a description for each code, an example of its correct use, and inclusion and exclusion criteria for use.

To achieve a high degree of internal validity of data coding, we conducted intercoder reliability checks among the coders as well as with the coding supervisor to achieve an internal reliability of 85%, a standard accepted in qualitative research (Carey, Morgan, & Oxtoby, 1996; MacQueen et al., 1998). For the first few coding assignments, all interviewers separately coded the same documents, and the coding was compared line by line to ascertain coding agreement. We considered that coding agreement had been reached when analysts coded the same text under the same code. Discrepancies between the coders were discussed and consensus was reached through team meetings. At the beginning of the coding process, we made intercoder reliability checks more frequently to improve the coding as well as identify emergent codes that did not appear to fit the initial coding scheme. We continued making intercoder reliability checks throughout the coding process until approximately a third of the data was subject to such measures and we were satisfied that a high degree of intercoder reliability (85%) had been established.

The qualitative results presented in this article are based on summaries of the coded data.

Results from Short Survey Data

We first profile the participants of the focus groups and interviews, based on their responses to the short survey, to provide a context for the qualitative results to be discussed in the next section of the article.

Responses to the short survey showed the female participants had a median age of 25 and the male participants, 25.5. As expected, African American and Caucasian participants had attained significantly higher levels of education than the Latinos, most of whom were recent immigrants. Whereas 79% had a high school education or more among the former two groups, among Latinos, only 48% had a high school education or more ($p < .01$, chi-square test).

Participants were asked if they currently had each of the following types of relationship: (a) "steady" relationship ("a husband/wife, steady boyfriend/girlfriend, or other serious relationship"); (b) "casual" relationship ("casual boyfriend/girlfriend, less than serious relationship, or one-night stands"); or (c) a "new" partner ("someone you have recently started having a relationship with, but you don't yet know if it will be serious or casual"). Participants could mark more than one answer. The majority of both male and female participants (70% of men and 77% of women) had a steady relationship, about one-fifth reported casual relationships (22% of men and 15% of women), and 10% reported that they had a new partner (13% of men and 7% of women). None of these differences between men and women was statistically significant.

Significantly more African Americans and Caucasians reported casual partners compared with Latinos (21% African Americans, 27% Caucasians, and 3% Latinos; $p < .05$). In addition, the majority of Latinos reported that they live with a partner (85%) compared with fewer than one half of African Americans and Caucasians (37% and 41%, respectively; $p < .01$).

The participants did not perceive themselves to be at high risk of being infected with an STI or HIV. The survey question for focus group participants did not specify a time frame for considering their risk. However, the in-depth interview participants were asked the question in the context of the next 2 years. Some 47% of the women and 62% of the men answered that they had "not much chance" of getting an STI, and 55% of women and 62% of men said they had "not much chance" of getting HIV. Only 11% to 16% of men and women thought they had a "strong chance" of getting an STI or HIV. The remainder responded they had "some chance." The differences between men and women in their perceptions of risk were not statistically significant.

In the short questionnaire, we also asked focus group and in-depth interview participants in separate questions about each relationship type whether "a woman

Table 3. Percentage of Participants Who Agree that a Woman Should Always Tell Partner She Is Using Vaginal Product to Protect Against STIs and HIV, by Gender

Relationship Type	Women	Men
Total Number of Respondents	68	64
Steady		
Strongly Agree	41.2	50.0
Agree	38.2	34.4
Disagree	16.2	10.9
Strongly Disagree	4.4	4.7
Total	100%	100%
Casual		
Strongly Agree	27.9	40.0
Agree	41.2	44.6
Disagree	25.0	12.3
Strongly Disagree	5.9	3.1
Total	100%	100%
New*		
Strongly Agree	27.9	43.1
Agree	23.5	32.3
Disagree	42.7	18.5
Strongly Disagree	5.9	6.1
Total	100%	100%

Note. *$p < .05$.
Relationship type categories are steady, casual, and new. Steady refers to a husband/wife, steady boyfriend/girlfriend, or other serious relationship. Casual refers to a casual boyfriend/girlfriend, other less than serious relationship, or one-night stands. New refers to someone you have recently started having a relationship with, but you don't yet know if it will be serious or casual.

should always tell her [partner type] if she is using a vaginal product [like a microbicide] for protection against STDs and HIV." Although the majority of both women and men thought a woman should always tell her partner, regardless of the type of relationship, more men than women thought so (Table 3). Moreover, whereas the percentage of women responding "strongly agree" or "agree" declined steadily from serious to casual to new relationships, the same percentage of men (84.4%, 84.6%) agreed for both steady and casual relationships. Even for new relationships, three-quarters of the men thought women should always tell their partners (whereas half the women felt this way; $p < .05$). As we shall discuss later, the in-depth interviews revealed reasons why men feel so strongly about this issue.

Results from Focus Group and Interview Data

We discuss, below, our findings on relationship issues that will likely influence use of a future topical microbicide. We, first, discuss emergent values about risk-reduction behavior according to relationship type and how normative perceptions of risk in different types of relationships create a paradox for communication about the need for adopting prevention behavior. We then consider a number of scenarios that participants presented as potential opportunities to talk about microbicide use with their partners and next discuss women's and men's views about covert use. We observe that the paradox for communication illustrates a potential barrier to acceptability of future microbicides in serious, committed relationships and, thus, engenders a need for covert use within such relationships. We also note that norms accepting risk-reduction behavior in casual relationships may lead to microbicide use strategies that are less problematic, although no less covert. We conclude by comparing women and men's views on whose responsibility it would be to ensure that microbicides were used.

Variation in Risk Perception and Behavior Among Relationship Types

Similar to other research on condom use at different stages of relationships (Frank, Poindexter, Cox, & Bateman, 1995; Woodsong & Koo, 1999), participants said that people are more likely to introduce a new STI protection method at the beginning of a sexual relationship. Most people in sexual encounters with new partners do not know the partners' sexual histories, nor do they have expectations of monogamy. They acknowledge some risk of exposure to an STI and, thus, some need for protection.

> Yes, I'm more likely to use with new relationships for a while until I feel like I could really trust them. (African American adult woman, interview)
> For me in a new relationship I'm going to use all precautions because I don't know where it's going . . . . (African American adult woman, interview)

Although respondents reported that even in new relationships it can be awkward to introduce condoms, fearing being embarrassed or appearing to be promiscuous, the general consensus was that using some form of protection was more likely in new relationships and one-night stands.

Both female and male participants in focus groups and interviews expressed an apparently emergent social value endorsing a woman's right to protect herself from STIs, particularly in a new and/or casual relationship. Although participants expressed this as a normative value, in their personal behaviors they may not act upon such norms. The reason may be that it is difficult to carry out the norm (as with many norms) and, perhaps, because the value is not yet well established. As indicated in both individual interview data and personal experiences reported in the focus group setting, people have a difficult time taking what they believe to be the best course of action for risk reduction. However, respondents reported that use of a protection product (condoms) in new and/or casual relationships has become an accepted norm. It is seen as a demonstration of personal responsibility for protecting one's self, and in some cases, for protecting partners.

"The first time he and I had sex, we used a condom because, well, we didn't know each other's histories." (Latina adult, interview).

Another woman said, "If you are a single young woman today, it is your responsibility to protect yourself. If you don't want to get pregnant, you can buy a box of condoms just like a man can." (Caucasian adult woman, interview).

Once a couple has been together for a while or the relationship becomes more serious, participants considered it commonplace to discontinue condom use. The timeframe for this "cautionary" period can vary greatly, from a few days to a few months. Some respondents admitted that perceptions of a "safe partner" might be a fantasy, acknowledging that although monogamy agreements generally cover both partners, one partner might be faithful, whereas the other is not. Many male and female participants discussed experience of infidelities in their own monogamous relationships, highlighting their understanding that even those in serious relationships should not consider themselves free of risk for acquiring an STI.

A woman who had been married and got HIV from her husband said, "That doesn't mean everybody is like that but I have lost my, I have lost trust in men for their credibility of being a one-woman man." (African American adult woman, focus group).

"I got hos and I got my wifey. That's how it is. Every brotha think like this. You got your main chick. You treat her different than anybody else . . . . I don't give a damn about them other girls." (African American teen male, interview)

A Paradox for Communication About Risk

Our data indicated that emergent acceptance of the need to discuss risk and risk-reduction strategies, and norms for meaningful sexual relationships present a paradox for individuals in serious relationships. These individuals are supposed to talk about the wide range of issues that potentially affect their relationship, including pregnancy intentions, sexual fidelity, and the possible need to take precautions against STI/HIV transmission. Both male and female participants generally agreed that women in more serious relationships have a greater responsibility to discuss pregnancy and the need for STI/HIV prevention than those with shorter-term, less serious, and non-monogamous partners.

However, almost all of the male (particularly African American) participants stated that discussions of STIs and HIV/AIDS rarely occur with steady partners. When queried about potential microbicide use, participants considered that a woman in a serious relationship should not need to use such a product, since a serious relationship implies a level of agreement, trust, and faithfulness. Participants felt that such trust is a cornerstone for serious relationships, and it should, ideally, eliminate the need for protection. They noted that talking about the need for protection could harm this trust and raise suspicions. Thus, we observe that the types of relationships in which people are considered to have more responsibility to talk openly about such topics as a need for protection are the very ones where such need should not exist.

> If you in a relationship, OK, relationships all spring back to love, honesty, compatibility, and communication OK, this my girl now, why I got to tell or ask her if she using something for, for STD. Where the hell she doing to catch it from? If we got a relationship, that means that's me and you. (African American adult male, focus group)

Specifically, participants felt that both men and women in established relationships would be concerned that talking about a protection product would imply infidelity and/or suspected infection of either partner. Several female participants stated that men would believe a woman was unfaithful if she introduced a microbicide product into a serious relationship, and almost all male participants held the same opinion.

> Yeah, like if this your boyfriend and you tell him I'm using this and y'all been to the doctor with each other and you know each other's status, then you know he gonna feel like oh, she must be seeing other dudes, you know what I'm saying? (African American teen female, focus group)
> I feel like you should be able to talk but the thing of it is, it's like in some marriages, me talking to my wife about condoms, then I'm telling her that I'm in the street. (African American adult male, focus group)

Even for less serious relationships, communication about the need to use a protection product is quite problematic, as the literature on condom use indicates (Woodsong & Koo, 1999; Hirsch, Higgins, Bentley, & Nathanson, 2002). Although participants reported that use of protection in new or casual relationships was the accepted norm, acknowledging a risk of acquiring an STI, some nevertheless felt that they would put a new or non-serious relationship in jeopardy if they suggested using a topical microbicide.

> [Product use] would probably end a casual relationship because of trust issues; the lack of trust and your opinion of him is less. (African American adult woman, interview)
> When a man and a woman get into a relationship right . . . you hope she ain't got nothing, but there are some thing you just don't come out and discuss like a business arrangement. (African American adult male, focus group).

Discussions of STIs could surface uncomfortable issues associated with past relationships, and this discussion in itself could potentially put the relationship at risk.

"That just might open doors to other things. She might want to know about your past relationships." (African American male teen, focus group)

Some couples may choose to forego discussions regarding sexual history and opt to visit the clinic periodically for STI screening. As one teen respondent said, "We never talk about nothing like that. We don't have to worry about it. We go to the clinic like every two months." (African American male teen, interview). In addition, participants reported that in any case it is still common to believe that a potential partner's STI status can be surmised by their outward appearance or lifestyle.

In most of the focus groups and in-depth interviews, when asked about a future vaginal protection product, participants said they believe that a woman in a steady relationship should openly discuss her reasons for wanting to use a product for STI prevention with her partner. At the same time, they acknowledge that women in such relationships should not need protection. With respect to new and casual partners, most men agreed that they would respect a woman who would initiate use of a female protection method. As a Caucasian male teen said in a focus group, "More power to her!" Again, this may be an emergent value that is not yet widely acted upon.

Participants believed that more couples do discuss the need to use methods for pregnancy prevention in comparison to STI prevention. Talking about the need for pregnancy prevention is less problematic than talking about STIs, as childbearing is long-term, whereas most STIs are considered to be treatable. "STDs are curable but a child is long-lasting. Guys are less interested in STDs than birth control because there's less risk." (African American female teen, interview)

Furthermore, pregnancy prevention may not be stigmatizing. When asked if there is a difference between talking to their partners about preventing pregnancy versus STIs, most participants said talking about pregnancies was easier and "normal." Talking about pregnancy prevention is potentially much more pleasant than talking about diseases and sexual history. We surmise that such discussions about birth control may provide some couples an opportunity to talk about the relationship and how children may fit into the relationship.

> Yes it's a big difference man. I mean because usually, when talking to people about birth control, it's normal. It's like, okay, that's like something that you're not afraid to talk about. Something that people are more open to talk about. Child birth, pregnancy and all that good stuff.... That's more so, the natural consequence of having sex. But when you start to discuss STD's and the problem that can come from having sex, it becomes an issue. (African American adult male, interview)

Nevertheless, pregnancy prevention may not be something that men want to talk about either, particularly in casual or new relationships. Many men felt that they would leave it to the woman to bring up the need to use protection and if she did not, they would assume that she was protecting herself from pregnancy. Similarly, most men felt that women had the responsibility for using birth control, and if a woman was using it, the man would be less likely to wear condoms because the threat of pregnancy was reduced, and STIs were not as worrisome.

Timing of Communication: When to Tell Partner About Microbicide Use

As noted above, our participants acknowledged the difficulty in having discussions about their risk for STIs/HIV, regardless of relationship type. They also pointed out that the circumstances and timing of informing a partner about one's intention to use a vaginal microbicide would likely influence its acceptability and use. Although opinions differed with respect to different relationship types, many respondents felt that a male partner should be told what a vaginal microbicide product is for and why a woman would feel she needs to use it. Most respondents agreed that a person who has decided to use a microbicide should tell her partner soon after her decision, using a straightforward approach. In most situations, men and women thought that saying that the doctor had given it to them, or had told them about the product, would add credibility to the introduction of the product. However, some participants also thought that rather than presenting a microbicide as a straightforward protection product, they could circumvent the negative implications of introducing protection by presenting the new product as a sex enhancer, which had the added side benefit of providing protection. (We discuss this again when we consider covert use.) In general, however, participants acknowledged that people are reluctant to take actions—including talking about protection—that could interfere with sexual arousal and intercourse, as well as the aura of romance that surrounds it. Thus, the timing of telling is important.

When asked about the best time for a woman to tell her partner about her use of a vaginal microbicide, participants suggested a wide range of possible scenarios, including the following:

- Inform early—Many said they would prefer to discuss a microbicide soon after the woman has decided to use it, or during a regular conversation time (e.g., over dinner), giving them both an opportunity to discuss concerns.

- Intimate/private moment—Some respondents indicated that bringing up prevention concerns would best be received when the two were sharing some intimate moments together, talking about their sex lives. Many men said they would be open to any discussion as long as it did not ruin the heat of the moment. "I don't want it to be, you know, like brought up right . . . right before you're about to get something [sex] 'cause I don't want to . . . wonder if that shit [is] working." (Caucasian adult male, focus group)

- Immediately before sex—Some men and women thought that right before sex would be the best time to inform a partner about intended microbicide use. They reasoned that this was a natural time to think about risk-reduction behavior but also that a man might be less likely to object after he is aroused. Some thought they could frame microbicides as a sex enhancer and, thus, bring it up "in the moment" as something new to add spice to the encounter.

- After sex—Some men and women felt that a man should be told after sex, so he would not make comments based on preconceived notions. If he noticed the product, and said something about sex feeling better or different, then they could talk about it. Although some men as well as women felt it is wrong to wait until afterwards to tell a partner, women were more likely to say it is the best time.

In general, men said they would agree to a woman's request to use a microbicide rather than turn down an opportunity to have sex, and further stated that in general, men should respect a woman's decision to protect herself. However, there was an expectation that a woman in an established relationship should not need to protect herself.

Covert Use

As mentioned earlier, much of the global interest in microbicides is based on the possibility that women could use them without the active participation of the partner and perhaps even without his knowledge. Thus, we explored this issue in some detail. Recall that women participating in this study were provided vaginal lubricants in the form of gels and suppositories (as proxies for microbicides) to use during intercourse before their interviews or focus groups. They were not advised whether or not to inform their partners before use, and in subsequent discussions or interviews, they were asked about their experiences in introducing the lubricants as well as their views on the potential for covert use. Some of the men in the focus groups and interviews were partners of the female participants, but others were not and therefore had not experienced intercourse using the lubricants. In both the focus groups and the in-depth interviews, men were shown a sample of the lubricant proxies and then asked similar questions about them.

In the focus groups and in-depth interviews, we asked participants under what circumstances and in what types of relationships people would likely use microbicides without informing their partner. Both men and women stated that, in principle, it should be acceptable to talk with partners about the use of an STI/HIV prevention product. Furthermore, most women felt that a woman should be able to use a product without being obliged to tell her partner. Participants acknowledged that a woman's decision to tell her partner about her prevention concerns was closely related to the type of relationship, the personality of the man, her relative power in the relationship, and her own level of trust with that partner. Taking these factors into account, many felt that covert use of a vaginal microbicide would be easier than having to cope with suspicions and accusations that would arise from raising the issue of protection.

Women's Views on Covert Use

The paradox involved in communicating about product use within serious relationships plays into perceptions about the appropriateness of using a vaginal microbicide without informing one's partner. As noted above, women in serious relationships are supposed to have less need for a protection product, and suggesting use of a microbicide within a committed relationship is expected to be problematic. This expectation will likely contribute to the need for covert use by women in this dilemma. Maintaining covert use over the long run in a serious relationship, however, may not be practical. It may be easier to resort to indirect covert use, as explained in a later section.

The apparent emergent norm endorsing women's right to protect themselves in casual relationships should eliminate the need to tell the casual partner:

> I think it all depends on what type of partner you have 'cause if you have just a sex partner, then, no, it ain't none of his business, but if it's like your boy friend then, now I'm saying y'all can talk to each other like that. (African American teen female, focus group).

However, adult female participants reported a variety of reasons why they would tell a casual partner about using a product. These include concerns that he may be at risk of, or actually experience, a side effect like a "rash or irritation," that they would feel bad "holding something back from [their partner]," particularly because he may find out anyway (seeing the product wrapper in the trash can, feeling a difference in the vagina, etc.). However, women gave a greater range of reasons why a woman in a casual relationship would not tell her partner. These include (in decreasing order of frequency) the following:

- fear that bringing up the topic of STI/HIV prevention would suggest to the partner that the woman had an STI (and conversely, covert use might keep a partner from wondering if she already had an STI);
- fear that bringing up the subject of prevention would arouse trust issues as mentioned above;
- fear of abuse or violence;
- fear that the product would not be 100% effective but the partner would not use a condom because she was using a microbicide;
- belief that the partner would not care whether or not the woman was using a product (especially among African American women); and
- belief that if she asked her partner, the partner would say "no" and then she wouldn't be able to use the product at all (particularly Latinas).

Indirect Covert Use

A number of women suggested that they would be more likely to use microbicides if they could tell their partner that the product was providing contraceptive protection, perhaps as a back-up method to their regular birth control method. They also saw a potential for introducing a vaginal product as something to enhance the sexual encounter, which had the added side benefit of providing protection. They could thus circumvent the negative implications of introducing protection. ("The people you wouldn't want to tell I'm going to use this [vaginal protection product]—you can just say well we are going to use this cool thing [fun, flavorful lubricant]"; Caucasian teen female, focus group). These examples of indirect covert use point to the potential importance of being able to obscure the purpose of product use, rather than attempting to hide it altogether. (It also points to the need to determine the contraceptive effectiveness of microbicides, a topic not discussed in this article.) For example, the two women below talked about the advantages of using microbicides as sex enhancers.

"... [I]f they had different scents and flavors, it would attract a younger crowd." (African American teen woman, interview).

"For married people, the flavor and the sensation, when you've been married 22 years, you have to spice things up." (Caucasian adult woman, interview).

Men's Views on Covert Use

When asked about potential covert use of a vaginal microbicides, men were generally opposed to it (consonant with Table 3). They were opposed particularly for serious relationships, in which they felt women needed to tell their partners to maintain trust.

Yes, if it's a serious relationship, she should tell before and explain the reason why. If you're supposed to be with just that one person, you need to explain why. If it's a casual relationship, you don't have to explain nothing—not obligated if not committed to the relationship. (Caucasian adult male, interview).

If I found out, then I couldn't trust her anymore. (African American adult male, interview)

Other reasons for opposition to covert use, regardless of type of relationship, were related to product characteristics—for example, concerns about the product's potential side effects for the male partner, its presence during oral sex, the confusion caused by finding the substances unexpectedly, or simply awkward moments during sex if they noticed the product on their penis:

... [Y]eah she should be straight up and tell you well, I'm using protection but it may have a side effect." (African American adult male, focus group)

Yes, it would be fine for her to tell me. ... if I were to notice something out of the ordinary in the fluid or the moisture, one would know it was because of the product and one wouldn't worry or be wondering if "what is it, is she sick, or not sick or what," you avoid that. (Latina adult, interview)

... I don't want surprises on my penis. (African American adult male, focus group)

Some also felt that a microbicide product should not be used covertly because it should protect both partners, and both partners need to know they were being protected. Women agreed that protection of the men as well as the women would increase acceptability and use. One male participant took a larger view and observed that a microbicide protecting only women could have the effect of reducing overall transmission of STIs (since fewer women would be infected).

However, there was also some support among men for covert use because it would avoid the embarrassing conversation about sexual history and STI/HIV risk and might even flatter him into thinking his partner was not concerned that he could be infected.

> It would be great to have this product the woman could use without the man knowing. If the woman is concerned that her partner has something, she can use it and he probably won't even know. That way he won't get offended 'cause he won't know there's anything there, it'd be all natural . . . . I might think she wasn't concerned about me having something . . . and I'm gonna feel good about myself. (Caucasian adult male, interview).

In fact, if the microbicide also protected against pregnancy, then there would be no need to talk about using it for STI protection, thus saving both partners from the hurt feelings that could arise from suggesting that one or both of them needed protection. In this way, some men would support the indirect covert use strategy, both partners taking the attitude that the microbicide was being used for pregnancy protection.

> All of it [protection against pregnancy, STIs and HIV] you don't want to get to the moment and then have to use too many things. You could lose the moment by then. It would save time and embarrassment. If it's for all of it, you don't really have to talk about what it's for. No one's feelings get hurt, but you get protected at the same time. (Caucasian adult male, interview)

Men's support of covert use also depended on the relationship type. Almost all of the men approved of the concept of vaginal microbicides and felt that women should be able to use them without disclosure but only in risky situations (e.g., partner being unfaithful, woman being unfaithful) and risky relationship types (e.g., one-night stands, casual sex, commercial sex).

Whose Responsibility?

Opinions varied about whether the woman or the man should take responsibility for using vaginal microbicides. Many women and men agreed that if a microbicide protected only one sex, then whomever it protected should be the one responsible for using it. Men were more emphatic about this than women, however. Some women and men felt that, if it protected both partners, both should share the responsibility, especially in a serious relationship. They thought that the decision should not be up to either one of the partners but should be a joint decision.

Others (both male and female) saw it as empowering for women to take responsibility for protecting themselves regardless of whether the partner was protecting himself.

"The woman [is responsible.] She needs to get it; that way if she doesn't feel right talking about it, she can protect herself whether he's protecting himself or not." (Caucasian adult male, interview).

Many women, particularly African Americans, who thought that it was the woman's responsibility, believed that was the case because women are more dependable, more responsible, will think ahead, and cannot expect the man to be thinking of the woman. A Caucasian female teenager echoed this sentiment in an interview: "It's more the woman who would make sure it gets used. When men are in the heat of the moment, they don't think about it."

However, others (especially Latinas) expressed resentment that women already had the responsibility for childbearing and preventing pregnancy, and vaginal microbicides could make STI/HIV protection also their responsibility. "My husband [should be responsible.] He should do it all. They don't know the pain of child birth, they should do this one" (Latina adult, interview).

On the other hand, a number of men were very distrustful of women and considered that even if a woman was using a microbicide, they would not be willing to leave it up to her and take her word that she was using the product. This was true particularly vis-à-vis promiscuous women. These men would continue to use condoms for risky sex, or they would make sure they saw her insert the microbicide. "She got to put the cream in, he should watch her do it . . . . Just to make sure she did it . . . because she might say she did it but didn't do it (African American adult male, interview).

Discussion and Conclusion

Female-initiated topical microbicides have the potential to save millions of lives from HIV/AIDS (Rockefeller Foundation n.d.-a) and their introduction can only be anticipated with great hope. However, concerns also exist about the future introduction of this method, including the likely level of efficacy of the early versions of microbicides, the varying numbers of lives that would be saved at various efficacy levels, the possible abandonment of more efficacious condoms in favor of the early microbicides (Foss, Vickerman, Heise, & Watts, 2003; Karmon, Potts, & Getz, 2003; Rockefeller Foundation, n.d.-a)—and, additionally, the need to ensure that the public accurately understands not only the benefits, but the limitations in protection offered by early microbicides and that popular knowledge of people becoming infected with HIV while using early microbicides will not lead to the rejection of later, more efficacious, generations of microbicides (Koo et al., 2002). We suggest that it is also advisable to consider that the advent of female-initiated topical microbicides is likely to introduce many layers of complexities into relationships between women and men.

Overall, male and female participants agreed that the possibility of covert use of microbicides would be beneficial, and the most important advantage of covert use in all relationship types would be that it could avoid the embarrassment and other difficulties that often keep people from discussing STIs and their sexual history. Overwhelmingly, people thought women should be able to use a microbicide to protect themselves. The observed paradox for women in committed relationships could lead to very carefully maintained covert use, since women in such relationships are not supposed to need an STI protection product, nor are they likely to be willing to risk the trust that is so important for such relationships. The observed emergent norm for acknowledging risk in casual relationships could lead to covert use of a different sort—women automatically using a microbicide without any discussion, yet somewhat secure in an assumption that use will be accepted by the casual partner, should he find out. Thus, variations in the type of sexual relationship will likely influence norms for acceptability and use.

Most participants perceived communicating with the partner about needing STI protection to be so difficult that to avoid it, some espoused subterfuge—not by using protection secretly but by introducing microbicides as a sex enhancer or as a pregnancy prevention method. Even for more casual relationships, such indirect covert use seemed appealing. Several participants suggested that the most effective way to get people to use microbicides would be, in effect, to coach people in indirect covert use: advertising microbicides primarily as sex enhancers or as vaginal lubricants with the added advantage of offering protection. The implications of such a strategy need to be explored, weighing its life-saving potential against its support of deception (albeit potentially agreeable to both partners) within the most intimate of relationships and the perpetuation of silence about sexual risk. Furthermore, another issue arises from the opposition of some men to covert or indirect covert use on the grounds of their need and right to know that these substances may cause side effects for them and are protecting them from STIs/HIV (if in fact, microbicides are developed to confer protection on both partners).

The introduction of female-initiated microbicides may add other complexities into relationship dynamics—complexities that revolve around women's and men's responsibilities for protection against pregnancy and STIs. Men often felt they had some responsibility for STI protection, but many men felt that using contraceptives to prevent pregnancy was the woman's charge. Some women expressed the concern that if microbicides protected against pregnancy as well as STIs and HIV, then some men may feel that they had no responsibility at all for either pregnancy of STI/HIV prevention. The result would be that this female method that was intended to empower women would place the entire burden of protection on them.

There was mistrust among men and women of one another. Men's experience with being deceived about women's use of birth control (being "trapped" by pregnancy) may explain the views of distrust expressed by some men about women using microbicides; they needed to see the women insert them before enjoying the benefit of "condomless" intercourse. Women voiced distrust of their partners, believing that their supposedly committed partners would be unfaithful and thus put them at risk but not take steps to protect them.

On the other hand, both women and men expressed concern and support of one another. Some men stated that if they were being promiscuous, they would not want their partners to get an STI because of their actions, and some expressed concern about the possible side effects of microbicides for their partners as well as themselves. Finally, some men (especially Latinos) were pleased that the

lubricating effects of microbicides could make sex more pleasant for their partners. On the other side, some women expressed deep concern that the microbicides
also protect their partners and not just themselves, and they valued the increase in
sexual pleasure that microbicides offered over condoms for their partners as well
as for themselves.

The advent of vaginal microbicides stands to bring both benefits and problems
into the sexual relationships of women and men. Understanding these dynamics
could help pave the way for a smooth introduction of this new and promising
female method of STI/HIV protection. Research is needed to provide this understanding. Both challenges and opportunities abound. The dynamics are both
complex and apparently changing, as norms about STI/HIV protection for casual
and new relationships seem to be evolving. Research needs to be conducted to
determine whether such norms and values are indeed emerging, and to investigate how such changing norms may affect the adoption of a new female-initiated
method of protection, and how the introduction of such a method may shape the
norms.

References

Alliance for Microbicide Development. (2003). *Home Page.* Retrieved July 9, 2003, from http://www.
    microbicide.org
Alliance for Microbicide Development. (2004). *Alliance for Microbicide Weekly News Digest,*
    5(9), Retrieved March 18, 2004, from http://www.microbicide.org/publications/digest/news.
    digest_vol5no09.pdf
Amaro, H. (1995). Love, sex, and power: Considering women's realities in HIV prevention. *American*
    Psychologist, 50(6), 437–447.
Anderson, J. E., Wilson, R., Doll, L., Jones, T. S., & Barker, P. (1999). Condom use and HIV risk
    behaviors among U.S. adults: Data from a national survey. *Family Planning Perspectives,*
    31(1), 24–28.
Annan, K. (as quoted in Alliance for Microbicide Development, 2004). *Alliance for Microbicide Weekly*
    News Digest, 5(9), Retrieved March 19, 2004, from http://www.microbicide.org/publications/
    digest/news.digest_vol5no09.pdf
Beadnell, B., Baker, S. A., Morrison, D. M., & Knox, K. (2000). HIV/STD risk factors for women
    with violent male partners. *Sex Roles, 42*(7–8), 661–689.
Bernard, H. R. (2000). *Social research methods.* Thousand Oaks, CA: Sage Publications.
Blanc, A. K. (2001). The effect of power in sexual relationships on sexual and reproductive health: An
    examination of the evidence. *Studies in Family Planning, 32*(3), 189–213.
Bogart, L. M., Cecil, H., & Pinkerton, S. D. (2000). Hispanic adults' beliefs, attitudes, and intentions
    regarding the female condom. *Journal of Behavioral Medicine, 23*(2), 181–206.
Bowleg, L., Belgrave, F. Z., & Reisen, C. A. (2000). Gender roles, power strategies, and precaution-
    ary sexual self-efficacy: Implications for Black and Latina women's HIV/AIDS protective
    behaviors. *Sex Roles, 42*(7–8), 613–635.
Carey, J., Morgan, M., & Oxtoby, M. (1996). Intercoder agreement in analysis of responses to open-
    ended interview questions: Examples from tuberculosis research. *Cultural Anthropology Meth-*
    ods, 8(3), 1–5.
Carroll, L. (1991). Gender, knowledge about AIDS, reported behavioral change, and the sexual behavior
    of college students. *Journal of American College Health, 40*, 5–12.
Castañeda, D. (2000). The close relationship context and HIV/AIDS risk reduction among Mexican
    Americans. *Sex Roles, 42*(7–8), 551–580.

Catania, J. A., Kegeles, S. M., & Coates, T. J. (1990). Towards an understanding of risk behavior: An AIDS risk reduction model (ARRM). *Health Education Quarterly, 17*(1), 53–72.

Centers for Disease Control and Prevention, National Center for HIV, STD and TB Prevention, Divisions of HIV/AIDS Prevention. (2003, March 27). *Fact sheet—HIV/AIDS among US women: Minority and young women at continuing risk* (rev. ed.). Retrieved August 12, 2003, from http://www.cdc.gov/hiv/pubs/facts/women.html

Christianson, M., Johansson, E., Emmelin, M., & Westman, G. (2003). "One-night stands"—risky trips between lust and trust: Qualitative interviews with Chlamydia trachomatis infected youth in north Sweden. *Scandinavian Journal of Public Health, 31*(1), 44–50.

Choi, K. H., Catania, J. A., & Dolcini, M. M. (1994). Extramarital sex and HIV risk behavior among US adults: Results from the National AIDS Behavioral Survey. *American Journal of Public Health, 84*(12), 2003–2007.

Choi, K. H., Gregorich, S. E., Anderson, K., Grinstead, O., & Gomez, C. A. (2003). Patterns and predictors of female condom use among ethnically diverse women attending family planning clinics. *Sexually Transmitted Diseases, 30*(1), 91–98.

Denzin, N. K., & Lincoln, Y. S. (Eds.). (2000). *Handbook of qualitative research* (2nd ed.). Thousand Oaks, CA: Sage Publications.

DiClemente, R. J. (1991). Predictors of HIV-preventive sexual behavior in a high-risk adolescent population: The influence of perceived peer norms and sexual communication on incarcerated adolescents' consistent use of condoms. *Journal of Adolescent Health, 12*, 385–390.

Ellen, J. M., Cahn, S., Eyre, S. L., & Boyer, C. B. (1996). Types of adolescent sexual relationships and associated perceptions about condom use. *Journal of Adolescent Health, 18*, 417–421.

Ellertson, C., & Burns, M. (2003). Re-examining the role of cervical barrier devices. *Outlook, 20*(2), 1–8.

Fisher, J., & Fisher, W. (2000). Theoretical approaches to individual-level behavior in HIV risk behavior. In J. Peterson & R. DiClemente (Eds.), *Handbook of HIV Prevention* (pp. 3–56). New York: Kluwer Academic/Plenum.

Ford, K., & Norris, A. E. (1993). Knowledge of AIDS transmission, risk behavior, and perceptions of risk among urban, low-income, African-American and Hispanic youth. *American Journal of Preventive Medicine, 9*, 297–306.

Ford, K., & Norris, A. E. (1995). Factors related to condom use with casual partners among urban African-American and Hispanic males. *AIDS Education and Prevention, 7*, 494–503.

Foss, A. M., Vickerman, P. T., Heise, I., & Watts, C. H. (2003). Shifts in condom use following microbicide introduction: Should we be concerned? *AIDS, 17*(8), 1227–1237.

Frank, M. L., Poindexter, A. N., Cox, C. A., & Bateman, L. (1995). A cross-sectional survey of condom use in conjunction with other contraceptive methods. *Women Health, 23*, 31–46.

Glaser, B., & Strauss, A. (1967). *The discovery of grounded theory: Strategies for qualitative research.* New York: Aldine.

Gutierrez, L., Oh, H. J., & Gillmore, M. R. (2000). Toward an understanding of (em)power(ment) for HIV/AIDS prevention with adolescent women. *Sex Roles, 42*(7–8), 581–611.

Harvey, S. M. (2001). Preventing HIV/STDs and unintended pregnancies: A decade of challenges. *Population and Environmental Psychology Bulletin, 27*(3), 1–6.

Harvey, S. M., Beckman, L. J., Browner, C. H., & Sherman, C. A. (2002). Relationship power, decision making, and sexual relations: An exploratory study with couples of Mexican origin. *Journal of Sex Research, 39*(4), 284–291.

Harvey, S. M., Bird, S. T., & Branch, M. R. (2003). A new look at an old method: The diaphragm. *Perspectives on Sexual and Reproductive Health, 35*(6), 270–273.

Harvey, S. M., Bird, S. T., Galavotti, C., Duncan, E. A. W., & Greenberg, D. (2002). Relationship power, sexual decision making and condom use among women at risk for HIV/STDs. *Women and Health, 36*(4), 69–84.

Hingson, R., Strunin, L., & Berlin, B. (1990). Acquired immunodeficiency syndrome transmission: Changes in knowledge and behaviors among teenagers, Massachusetts statewide surveys, 1986 to 1988. *Pediatrics, 85*(1), 24–29.

Hingson, R. W., Strunin, L., Berlin, B. M., & Heeren, T. (1990). Beliefs about AIDS, use of alcohol and drugs, and unprotected sex among Massachusetts adolescents. *American Journal of Public Health, 80*, 295–299.

Hirky, A. E., Kirshenbaum, S. B., Melendez, R. M., Rollet, C., Perkins, S. L., & Smith, R. A. (2003). The female condom: Attitudes and experiences among HIV-positive heterosexual women and men. *Women and Health, 37*(1), 71–89.

Hirsch, J. S., Higgins, J., Bentley, M. E., & Nathanson, C. A. (2002). The social constructions of sexuality: Martial infidelity and sexually transmitted disease-HIV risk in a Mexican migrant community. *American Journal of Public Health, 92*(8), 1227–1237.

Impett, E. A., & Peplau, L. A. (2003). Sexual compliance: Gender, motivational, and relationship perspectives. *Journal of Sex Research, 40*, 87–100.

Jenkins, S. R. (2000). Toward theory development and measure evolution for studying women's relationships and HIV infection. *Sex Roles: A Journal of Research, 42*(7–8), 751–780.

Jewkes, R. K., Levin, J. B., & Penn-Kekana, L. A. (2003). Gender inequalities, intimate partner violence and HIV preventive practices: Findings of a South African cross-sectional study. *Social Science & Medicine, 56*(1), 125–134.

Karim, Q. A., Karim, S. S. A., Soldan, K., & Zondi, M. (1995). Reducing the risk of HIV-infection among South-African sex workers—socioeconomic and gender barriers. *American Journal of Public Health, 85*(11), 1521–1525.

Karmon, E., Potts, M., & Getz, W. M. (2003). Microbicides and HIV: Help or hindrance? *Journal of Acquired Immune Deficiency Syndrome, 34*(1), 71–75.

Koo, H. P., Woodsong, C., Simons-Rudolph, A., Dalberth, B., Koch, M. A., Viswanathan, M., et al. (2002, November). *Role of individual and sociocultural factors and product characteristics in the likely use of vaginal microbicides.* Paper presented at the annual meeting of the American Public Health Association, Philadelphia, PA.

Krueger, A. A. (1994). *Focus groups: A practical guide for applied research.* Thousand Oaks, CA: Sage Publications.

Lauby, J. L., Semaan, S., O'Connell, A., Person, B., & Vogel, A. (2001). Factors related to self-efficacy for use of condoms and birth control among women at risk for HIV infection. *Women & Health, 34*(3), 71–91.

Logan, T. K., Cole, J., & Leukefeld, C. (2002). Women, sex, and HIV: Social and contextual factors, meta-analysis of published interventions, and implications for practice and research. *Psychological Bulletin, 128*(6), 851–885.

Luke, N. (2003). Age and economic asymmetries in the sexual relationships of adolescent girls in sub-Saharan Africa. *Studies in Family Planning, 34*(2), 67–86.

Macaluso, M., Demand, M., Artz, L., Fleenor, M., Robey, L., Kelaghan, J., & Cabral, R., Hook, E. W., III. (2000). Female condom use among women at high risk of sexually transmitted disease. *Family Planning Perspective, 32*(3), 138–144.

MacQueen, K. M., McLellan, E., Kay, K., & Milstein, B. (1998). Codebook development for team-based qualitative analysis. *Cultural Anthropology Methods, 10*(12), 31–36.

Malow, R. (2001). Topical microbicides: Removing barriers to HIV prevention. *STEP Ezine, 1*(25), 1–5.

Miles, M., & Huberman, A. M. (2000). *Qualitative data analysis: An expanded sourcebook* (2nd ed.). Thousand Oaks, CA: Sage Publications.

Mill, J. E., & Anarfi, J. K. (2002). HIV risk environment for Ghanaian women: Challenges to prevention. *Social Science & Medicine, 54*(3), 325–337.

Miller, M., & Neaigus, A. (2001). Networks, resources and risk among women who use drugs. *Social Science & Medicine, 52*(6), 967–978.

Morgan, D. (Ed.). (1993). *Successful focus groups: Advancing the state of the art.* Newbury Park, CA: Sage Publications.

Nathanson, C. A., Upchurch, D. M., Weisman, C. S., Kim, Y. J., Gehret, J., & Hook, E. W., III. (1993). *Sexual risks, sexual relationships, and sexual behavior among African-American male STD clinic clients* (Paper on Population WP 93-01). Baltimore, MD: Johns Hopkins Population Center.

National Institute of Allergy and Infectious Diseases. (2000). *NIAID topical microbicide research: Developing new tools to protect women from HIV/AIDS and other STDs.* Washington, DC: Author.

O'Leary, A. (2000). Women at risk for HIV from a primary partner: Balancing risk and intimacy. *Annual Review of Sex Research, 11*, 191–235.

Overby, K. J., & Kegeles, S. M. (1994). The impact of AIDS on an urban population of high-risk female minority adolescents: Implications for intervention. *Journal of Adolescent Health, 15*, 216–227.

Parker, R. (2001). Sexuality, culture and power in HIV/AIDS research. *Annual Review of Anthropology, 30*, 163–179.

Parker, R. G., Easton, D., & Klein, C. H. (2000). Structural barriers and facilitators in HIV prevention: A review of international research. *AIDS, 14*(Suppl. 1), S22–S32.

Plichta, S. B., Weisman, C. S., Nathanson, C. A., Ensminger, M. E., & Robinson, J. C. (1992). Partner-specific condom use among adolescent women clients of a family planning clinic. *Journal of Adolescent Health, 13*, 506–511.

Population Council. (2001). *The case for microbicides: A global priority.* Second Edition. New York: Author.

Powell, M. G., Mears, B. J., Deber, R. B., & Ferguson, D. (1986). Contraception with the cervical cap: Effectiveness, safety, continuity of use, and user satisfaction. *Contraception, 33*(3), 215–232.

Prochaska, J. O., Redding, C. A., Harlow, L. L., Rossi, J. S., & Velicer, W. F. (1994). The transtheoretical model of change and HIV prevention: A review. *Health Education Quarterly, 21*(4), 471–486.

Pulerwitz, J., Amaro, H., De Jong, W., Gortmaker, S. L., & Rudd, R. (2002). Relationship power, condom use and HIV risk among women in the USA. *AIDS Care-Psychological and Socio-Medical Aspects of AIDS/HIV, 14*(6), 789–800.

Pulerwitz, J., Gortmaker, S. L., & DeJong, W. (2000). Measuring sexual relationship power in HIV/STD research. *Sex Roles, 42*(7–8), 637–660.

Quina, K., Harlow, L. L., Morokoff, P. J., & Burkholder, G. (2000). Sexual communications in relationships: When words speak louder than actions. *Sex Roles, 42*(7–8), 523–549.

Reid, P. T. (2000). Women, ethnicity, and AIDS: What's love got to do with it? *Sex Roles, 42*(7–8), 709–722.

Richards, T. (2002). *NVivo 2.0.* Victoria, Australia: QSR International.

Rockefeller Foundation. (n.d.-a). *The public health benefits of microbicides in lower-income countries: Model projections. A report by the Public Health Working Group of the Microbicide Initiative.* New York, NY: Rockefeller Foundation. Retrieved May 6, 2004, from http://www.rockfound. org/Documents/488/rep7_publichealth.pdf

Rockefeller Foundation. (n.d.-b). *The science of microbicides: Accelerating development. A report by the Science Working Group of the Microbicide Initiative.* New York: Author. Retrieved August 29, 2003, from http://www.microbicide.org/microbicideinfo/rockefeller/science.of. microbicides.rockfound.pdf

Sacco, W. P., Rickman, R. L., Thompson, K., Levine, B., & Reed, D. L. (1993). Gender differences in AIDS-relevant condom attitudes and condom use. *AIDS Education and Prevention, 5*, 311–326.

Santelli, J. S., Kouzis, A. C., Hoover, D. R., Polacsek, M., Burwell, L. G., & Celentano, D. D. (1996). Stage of behavior change for condom use: The influence of partner type, relationship and pregnancy factors. *Family Planning Perspectives, 28*(3), 101–107.

Seage, G. R., Holte, S., Gross, M., Koblin, B., Marmor, M., Mayer, K. H., & Lenderking, W. R. (2002). Case-crossover study of partner and situational factors for unprotected sex. *Journal of Acquired Immune Deficiency Syndromes, 31*(4), 432–439.

Seidman, I. E. (1991). *Interviewing as scientific research.* New York, NY: Teachers College Press.

Sherman, S. G., Gielen, A. C., & McDonnell, K. A. (2000). Power and attitudes in relationships (PAIR) among a sample of low-income, African-American women: Implications for HIV/AIDS prevention. (Statistical data included). *Sex Roles: A Journal of Research, 42*(3–4), 283–294.

Simoni, J. M., Walters, K. L., & Nero, D. K. (2000). Safer sex among HIV+ women: The role of relationships. *Sex Roles, 42*(7–8), 691–708.

Spradley, J. (1979). *The ethnographic interview.* New York: Holt, Rinehart and Winston.

Strauss, A., & Corbin, J. (1990). *Basics of qualitative research: Grounded theory procedures and techniques.* Newbury Park, CA: Sage Publications.

Tang, C. S. K., Wong, C. Y., & Lee, A. M. (2001). Gender-related psychosocial and cultural factors associated with condom use among Chinese married women. *AIDS Education and Prevention, 13*(4), 329–342.

Ulin, P. R., Robinson, E. T., Tolley, E. E., & McNeill, E. T. (2004). *Qualitative Methods: A field guide for applied research in sexual and reproductive health* (2nd ed.). San Francisco, CA: Jossey Bass.

Vanwesenbeeck, I., van Zessen, G., Ingham, R., Jaramazovic, E., & Stevens, D. (1999). Factors and processes in heterosexual competence and risk: an integrated review of the evidence. *Psychology and Health, 14*, 25–50.

Wingood, G. M., & DiClemente, R. J. (2000). Application of the theory of gender and power to examine HIV-related exposures, risk factors, and effective interventions for women. *Health Education & Behavior, 27*(5), 539–565.

Woodsong, C., & Koo, H. (1999). Two good reasons: Women's and men's perspectives on dual contraceptive use. *Social Science and Medicine, 49*, 567–580.

HELEN P. KOO, is a Senior Research Demographer at RTI International. She obtained her DrPH in Population Planning from the University of Michigan in 1973. Since then, Dr. Koo has conducted research and evaluation in a variety of subjects, including most recently unintended pregnancy and contraceptive behavior, adolescent pregnancy prevention, and acceptability of long-acting contraceptives. She is the principal investigator of the study reported in this article, and also of the methodologically related study of unintended pregnancy, which, like the reported study, integrated qualitative and quantitative methods. She has also directed research grants focusing on fertility timing patterns, infant mortality, marital disruption and remarriage, and family and household structure.

CYNTHIA WOODSONG earned a PhD in Anthropology from the State University of New York in 1992 and completed a postdoctoral training program at the Carolina Population Center, University of North Carolina in 1994. Dr. Woodsong is a Senior Scientist at Family Health International (FHI) in Durham, North Carolina. At FHI, she is working on aspects of use-adherence and acceptability of microbicides in clinical trial settings, including work with the HIV Prevention Trials Network (HPTN). She is an active member of the Ethics Workgroup of HPTN, having developed an expanded model of the informed consent process, and currently chairs an HPTN sub-committee that is identifying and developing ethics training resources for HIV prevention research. She is also currently principal investigator on a study of the validity of self-report behavior, being conducted in Zimbabwe among a population expected to participate in microbicides clinical trials. Working with Dr. Koo, she co-directed the qualitative phase of a study of unintended pregnancy in the United States and directed the qualitative phase of the study that is reported on in this issue.

BARBARA T. DALBERTH is a Public Health and Policy Analyst at RTI International with over 10 years of public health research experience. She obtained her MPH from the University of North Carolina in 1996. Ms. Dalberth has worked extensively in research on family planning and sexually transmitted infection (STI) prevention and child welfare issues.

MEERA VISWANATHAN is a Research Health Analyst at RTI International. She obtained her PhD from the University of North Carolina at Chapel Hill in 2001. Dr. Viswanathan's research interests center on methodological and substantive issues in child and maternal health outcomes. She is involved with multiple evidence-based reviews on clinical and methodological topics, including the epidemiology of HIV.

ASHLEY SIMONS-RUDOLPH is a Health Analyst at RTI International and a doctoral candidate at The George Washington University. She holds a Master's degree in Public Policy from The George Washington University and a BA in Psychology from North Carolina State University. Her previous publications have focused on the dissemination of public policy, qualitative methods, school-based substance use prevention, and welfare policy.

94

Journal of Social Issues, Vol. 61, No. 1, 2005, pp. 95–107

Development and Evaluation of the Abortion Attributes Questionnaire

S. Marie Harvey*

University of Oregon

Mark D. Nichols

Oregon Health Sciences University

This paper describes the development and evaluation of the Abortion Attributes Questionnaire (AAQ), an instrument designed to assess the perceived importance of specific characteristics of abortion methods. Women receiving medical abortions (n = 186) and women who chose surgical procedures (n = 118) completed the AAQ. Participants were asked to rate how important each of 21 characteristics would be "when choosing between surgical and medical abortion." Factor analyses revealed that the AAQ consists of four factors and, as subscales, the factors have good internal reliability. The validity of the AAQ was established through discriminant function analysis and results indicated that three of the four factors predicted choice. Taken together, these findings provide empirical support for the constructs measured by the AAQ.

Women in the United States seeking to terminate an early pregnancy now have a choice between medical and surgical abortion procedures. Medical abortion is the termination of pregnancy using a drug or combination of drugs that can be administered orally, intramuscularly, and/or vaginally, first causing the

---

*Correspondence concerning this article should be addressed to S. Marie Harvey, Center for the Study of Women in Society, 1201 University of Oregon, Eugene, OR 97403-1201 [e-mail: mharvey@oregon.uoregon.edu].

We extend our appreciation to the site coordinators and staff at the participating Planned Parenthood clinics, especially Sue Ferden in Iowa, Madge Armstrong in Pennsylvania, Evonne Morici and Rebecca Whiteman in California, Jini Tanenhaus in New York, and Vicki Jacobs in Arizona. We thank Linda J. Beckman and Sarah J. Satre for their contributions to this study and Anna Wilson for her assistance with data analysis. Funding for this research was provided by grants from The John Merck Fund and the Education Foundation of America to the Public Health Institute. We heartily thank them for their generous support. Finally, we thank the women who took the time to share their perceptions and experiences with us.

pregnancy to terminate and then causing the uterus to expel the products of conception. In comparison to medical abortion, surgical abortion involves inserting an instrument through the cervix into the uterus and removing the products of conception. On September 28, 2000, after 12 years of political struggle, the U.S. Food and Drug Administration (FDA) gave final approval to release mifepristone (commonly known as RU 486 or the "French abortion pill") for use as an abortifacient. Because antiprogestins like mifepristone were unavailable in the United States until recently, medical researchers in the early 1990s documented the safety and effectiveness of another drug, methotrexate, to induce medical abortion (Creinin, 1993; Creinin & Darney, 1993; Hausknecht, 1995; Schaff, Eisinger, Franks, & Kim, 1995). Because methotrexate has had FDA approval since 1953 for the treatment of cancer, its off-label use allows physicians to legally prescribe this drug for medical abortion. In the first half of 2001, providers performed approximately 37,000 medical abortions using both methotrexate and mifepristone, representing 6% of all abortions during that period (Finer & Henshaw, 2003).

The availability of medical abortifacients holds the promise of increasing access to abortion for women in the United States. This new technology will, however, increase options and access to abortion only if women in the United States find it acceptable and use it. The efficacy of medical abortion is highly dependent upon a woman's willingness to adhere to the regimen and to wait for the drugs to take action. Thus, the success of this method is inextricably linked to its acceptability. The development of an instrument that would assist health care providers in determining which characteristics of abortion methods are important to a woman would be helpful in matching a woman with a method that she finds acceptable.

Presumably, the process of choosing an abortion method is strongly influenced by the perceived attributes of the specific methods. Several international studies have examined why women choose a particular abortion procedure, especially medical abortion. Findings from these studies indicate that choice is often attributed to desiring specific characteristics of the method (summarized in Winikoff, 1995). Despite the diversity of research designs of the published studies that have examined factors that influence choice, the findings are quite consistent. Fear of surgery and general anesthesia were consistently identified in these studies as reasons for choosing medical abortion (Berer, 1992; Tang, 1991; Thong, Dwar, & Baird, 1992). In addition, women reported that they chose medical abortion because it seemed more natural, like a premeditated miscarriage, and allowed for more privacy (Berer, 1992; Tang, 1991; Thong et al., 1992). Multiple visits, greater length of time and inconvenience (Bachelot, Cludy, & Spira, 1992; Tang, 1991; Tang, Lau, & Yip, 1993), as well as greater pain and bleeding experienced (Rosen, Von Knorring, & Bygdeman, 1984; Tang et al., 1993) were mentioned as disadvantages of the method as compared to surgical abortion.

Although several international studies have examined why women choose a particular abortion procedure, especially medical abortion, very little is known about factors that influence choice of abortion method among women in the United States. Results from studies of U.S. women (e.g., Beckman & Harvey, 1997; Creinin & Park, 1995; Elul, Pearlman, Sohaindo, Simonds, & Westhoff, 2000; Fielding, Edmunds, & Schaff, 2002; Winikoff, Ellertson, Elul, & Sivin, 1998) parallel those of international studies and indicate that women choose medical abortion for several reasons including the following: the method allowed them to avoid surgery (Beckman & Harvey, 1997; Creinin & Park, 1995; Elul et al., 2000; Fielding et al., 2002; Winikoff et al., 1998); they felt the method was safer than vacuum aspiration (Winikoff et al., 1998); they believed that medical abortion was a more natural method (Beckman & Harvey, 1997; Elul et al., 2000; Fielding et al., 2002; Winikoff et al., 1998); they felt that this method allowed for the least risk of infection (Beckman & Harvey, 1997); the method could be used early in pregnancy (Beckman & Harvey, 1997) and the method allowed them more privacy and autonomy (Elul et al., 2000; Fielding et al., 2002; Winikoff et al., 1998). Taken together, findings from international and U.S. studies provide evidence that specific attributes are related to method desirability and actual use.

To better understand the influence of perceived attributes on choice and actual use of abortion methods, these attributes need to be accurately assessed using a reliable and valid research instrument. This article describes the development and evaluation of the Abortion Attributes Questionnaire (AAQ), a research tool designed to assess the perceived importance of specific characteristics of abortion methods. Evidence supporting the reliability and validity is presented.

Methods

Participants

This study was conducted in conjunction with the Planned Parenthood Federation of America (PPFA) clinical trial to evaluate the medical abortion experience for women and determine the effectiveness of methotrexate and misoprostol for early term abortions. Our study supplemented the larger study and focused on the acceptability and user-perspective of medical-induced abortion. The sample for this study, a subgroup of the larger trial, included 186 women who had methotrexate-induced abortions while enrolled in the PPFA clinical trial at one of five clinics: Des Moines, Iowa; York, Pennsylvania; Phoenix, Arizona; Walnut Creek, California; and New York, New York. The comparison sample included 118 women who were eligible for the clinical trial and were offered the option of a medical abortion but chose instead the surgical procedure.

From September 1997 through July 1998 clinic personnel at the participating PPFA affiliates recruited all women who were 18 years and older and met the eligibility requirements for the methotrexate clinical trial to participate in this acceptability study.

These women were given written information about the acceptability study after they had been counseled on abortion method options and had chosen either medical or surgical abortion. Clinic personnel explained to the women that their participation was completely voluntary and would not affect the care they receive at Planned Parenthood. If a woman expressed interest, written consent for the acceptability study was obtained and brief instructions on how to complete the questionnaires were given.

Of the 368 eligible women asked to participate, 304 (82.6%) consented and completed the questionnaire (80.8% of surgical group, 86.1% of medical group). Women who agreed to participate were compared to those who refused on three demographic characteristics—age, ethnicity and education. Although the two groups did not significantly differ on age and ethnicity, women who participated in the study reported significantly higher levels of education compared to non-participants ($t = 2.60$, $p < .05$).

Data Collection

Data were collected using a pre-tested, self-administered questionnaire. Women completed the questionnaire after they had chosen their abortion method but before the abortion was initiated (i.e., before the injection was given or the surgical procedure performed). The instrument collected background information and data on the following topics: abortion method chosen; expectations of the method chosen—with regard to pain, anxiety, and bleeding; and the relative importance of method characteristics in their decision-making. Clinic staff verified the methods reported by women.

Measure

The attributes included in the AAQ were derived from two sources: (a) previous studies that examined reasons women reported for choosing an abortion method, particularly medical abortion (e.g., Beckman & Harvey, 1997; Creinin & Park, 1995; Winikoff et al., 1998) and (b) focus group findings from a qualitative study that examine perceived advantages and disadvantages of medical abortion compared to surgical abortion (Harvey, Beckman, Castle, & Coeytaux, 1995). The final version of the AAQ consisted of 21 characteristics of abortion methods (see Appendix).

To determine the characteristics women value in an abortion method, we asked participants to rate how important each of 21 characteristics would be "when choosing between surgical and medical abortion." Participants recorded their responses

on a 5 point Likert scale that ranged from a value of 1 "not important" to a value
of 5 "extremely important."

Analysis

We generated descriptive statistics (frequencies and means) for all variables
of interest. Bivariate analyses were conducted to compare the medical and surgical
groups on demographic and background characteristics. We compared categorical
variables using chi-square analysis and the continuous variable (mean age) using a
t-test. Because of the small numbers in cells, we collapsed some variables into fewer
categories. More specifically, we dichotomized race/ethnicity into non-Hispanic
White versus all other racial/ethnic groups. Similarly, we collapsed comfort with
the decision to have an abortion into two categories: (a) very comfortable and
(b) somewhat comfortable, somewhat uncomfortable, and very uncomfortable.

To examine the structure of the AAQ and to determine if the scale included
more than one domain, we conducted principal components factor analysis with
varimax rotation on the set of 21 importance items. All respondents that were
missing more than one-third of the 21 items were eliminated from the analyses ($N =$
6 or 2%). In addition, for participants missing fewer than one-third of the items, the
mean for the missing item was replaced with the mean of the sample for that item.
Internal consistency reliability of the factors were ascertained using Cronbach's
alpha. To assess the construct validity of the AAQ we performed a discriminant
analysis using stepwise procedures to determine if the factors predicted whether
women chose medical or surgical abortions.

Results

Demographic and Background Characteristics

Women in the medical and surgical abortion groups did not significantly differ
on any of the demographic characteristics examined (Table 1). Ages of participants
in both groups ranged from 18 to 45 with a mean age of 27.5 years. Over two-
thirds of the women were non-Hispanic White, 12.4% were African American, and
11.0% were Hispanic/Latina. Although not statistically significant, more women
in the medical group (40.3%) than in the surgical group (27.9%) had completed
college. Three-fourths of the participants were single (76.2%) and had jobs outside
the home (77.4%).

As with demographic characteristics, the two samples were similar with regard to other background variables. Half of the women (51.5%) reported having
children. One in five women had experienced a miscarriage (20.5%) and almost
half (47.4%) had experienced a previous surgical abortion. Most women (84.4%)
reported feeling very comfortable or somewhat comfortable with their decision to

Table 1. Mean Values, Percentage Distribution of Abortion Patients, by Selected Characteristics, and Percentages with Certain Characteristics, all According to Abortion Method Chosen

Characteristic	Total (N = 304)	Surgical (N = 118)	Medical (N = 186)
Age (mean)	27.5	27.2	27.6
% Non-Hispanic White	68.2	73.0	65.2
Education			
High school	23.1	26.2	21.0
Some college	41.4	45.8	38.7
Completed college/Some graduate	35.5	27.9	40.3
Marital Status			
Married	23.8	21.2	25.4
Not married	76.2	78.8	74.6
Employment Status			
Not working outside home	22.6	21.2	23.5
Working full-time	53.8	56.8	51.9
Working part-time	23.6	22.0	24.6
Previous Births			
None	48.5	49.6	47.8
1 or more	51.5	50.4	52.2
% ever miscarried	20.5	18.6	21.6
% ever had an abortion	47.4	42.4	50.5
% very comfortable with decision to have abortion	44.2	42.2	45.4
% ever raped/sexually abused	22.1	19.8	23.5

Note. The two abortion groups did not significantly differ on any of these characteristics.

have an abortion. More than one-fifth of the participants (22.1%) reported having been raped, sexually abused, or forced to have sex at some time in their lives.

Dimensions of the AAQ

Principal components factor analysis with varimax rotation was performed on the set of importance items to examine whether the scale included more than one dimension. Because the scale was being developed and we did not have hypotheses about what factors to predict, exploratory factor analysis was conducted. Factors with eigenvalues greater than 1.0 were retained. Each factor consisted of the items loading at .50 or above on that factor. All items that received a factor loading of less than .50 were dropped from further analysis. Moreover, in order to purify the list, we eliminated items with loadings of .50 or greater on more than one factor. Five items were deleted and another factor analysis was conducted with the 16 items that were retained.

The factor analysis of ratings on these 16 items produced four factors that were rotated to terminal solution (Table 2). The resulting four factors explained 65.2% of the variance in the importance ratings. Factor 1 accounted for 22.9% of the variance and was labeled absence of side effects. This factor consists of six items that describe minor side effects associated with abortion procedures.

Table 2. Rotated Factor Loadings for the Four Dimensions of the Abortion Attributes Questionnaire

Attribute	Factor 1	Factor 2	Factor 3	Factor 4
doesn't involve painful cramping for more than an hour	.816			
doesn't involve heavy bleeding	.790			
doesn't involve bleeding for more than 7 days	.789			
doesn't involve seeing the blood and blood clots during the abortion	.750			
doesn't have side effects like nausea, headache, and diarrhea	.737			
has been used for abortion for a long time	.598			
doesn't involve surgical instruments being inside you		.889		
lets you have abortion in the privacy of your own home		.843		
is like a natural miscarriage		.828		
doesn't involve surgery		.763		
is very effective			.744	
can be used very early in pregnancy			.725	
hardly ever causes major health problems (infection, tearing the uterus)			.647	
lets you have the abortion over with quickly			.632	
lets you have control during the abortion				.782
lets you know where and when the abortion is going to happen				.577

Factor 2, natural/avoidance of surgery, accounted for 19.2% of the variance. The four items that loaded on this factor concern avoiding surgical procedures, having the abortion away from the clinic in the privacy of one's home, and having an abortion that resembles a natural miscarriage. Factor 3, effectiveness and timing, accounted for 14.3% of the variance and consists of four items that focus on effectiveness, timing, and avoidance of major health problems. Factor 4, labeled personal control, accounted for 8.8% of the variance and consists of items that reflect having control during the abortion and letting you know when the abortion will occur. Each subscale score is the average of scores on the individual items that load on that factor, with higher scores indicating higher levels of importance for the factor.

Internal Consistency Reliability

We conducted several tests to evaluate the AAQ as a measurement tool. To assess the internal consistency reliability of the factors, Cronbach's alpha was calculated for each of the factors. The four factors, absence of side effects, natural/avoidance of surgery, effectiveness and timing, and personal control, achieved alphas of .87, .87, .66, and .59, respectively. Factors 1, 2, and 3 were deemed reliable because they met the minimum standard of reliability (Cronbach's alpha > .60; Nunnally & Bernstein, 1994). The lower reliability of Factor 4, with only two items, was expected because of the sensitivity of coefficient alpha to number of items. All four factors were, therefore, retained. The correlations among the four factors showed that the Factor 2, natural/avoidance of surgery, had the lowest correlation with the other factors ($rs = .07 - .29$). The highest correlation was between Factor 1, absence of side effects, and Factor 4, personal control ($r = .40$).

Construct Validity

As previously discussed, our goal was to develop an instrument that would discriminate between women who choose medical abortion and those who choose surgical abortion. To assess the construct validity of the AAQ we preformed a discriminant function analysis using stepwise procedures to determine the extent to which the four factors could predict choice of abortion method. The binary grouping variable or dependent variable was medical abortion users/surgical abortion users. The discriminant function was significant ($r = .72$, $x^2 = 217.21$, $df = 3$, $p < .001$). Three of the four factors predicted choice of abortion method (Table 3). In order of importance in terms of successful prediction were Factor 2—natural/avoidance of surgery, Factor 1—absence of side effects, and Factor 4—personal control. These findings indicate that women who chose medical abortion were more likely to place greater importance on a method not involving surgery, allowing the abortion to take place in the privacy of their home, and resembling a natural miscarriage. In contrast, women who chose the surgical procedure placed more importance on an abortion method that does not involve minor side effects (e.g., painful cramping for more than an hour, bleeding longer than 7 days) and allows them more personal control during the abortion procedure including knowing where and when the abortion will happen.

To test the adequacy of the of the discriminant function, jackknifed classification analysis was conducted. The percent correctly classified was 85.2%, 84.0% of the medical abortion users and 87.2% of the surgical abortion users. Thus, the findings demonstrate considerable agreement between predicted and actual group membership and suggest that the underlying factors of the AAQ are useful as predictive variables for choice of abortion method. These findings provide empirical support for the constructs measured by the AAQ.

Table 3. Summary Table of the Significant Predictor Variables in the Discriminant Function Analysis Predicting Choice of Abortion Method

Factor	Variable In	Standardized Coefficient	Wilk's Lambda
2. Natural/Avoidance of Surgery	1	−.95	.83
1. Absence of Side Effects	2	.73	.64
4. Personal Control	3	.26	.50

Note. Positive coefficient indicates greater importance among surgical abortion users. Negative coefficient indicates greater importance among medical abortion users.

Discussion

In this study we describe the development and evaluation of the Abortion Attributes Questionnaire (AAQ), an instrument designed to assess the perceived importance of specific characteristics of abortion methods. We developed the questionnaire by reviewing the literature that examined reasons women report for choosing an abortion method and women's reports of perceived advantages and disadvantages of medical and surgical abortion. We administered the questionnaire to a sample of women who chose either a medical or surgical abortion procedure for pregnancy termination. A factor analysis yielded four distinct factors which reflect salient characteristics of abortion methods: absence of side effects; natural/avoidance of surgery; effectiveness and timing; and personal control. The AAQ possesses good internal consistency reliability and demonstrates construct validity.

Findings from the discriminant function analysis support the construct validity of the AAQ with three of the four factors predicting choice of abortion method. Consistent with findings from earlier studies (e.g., Beckman & Harvey, 1997; Creinin & Park, 1995; Elul et al., 2000; Fielding et al., 2002; Winikoff et al., 1998), women who chose medical abortion were more likely to place greater importance on a method not involving surgery, allowing the abortion to take place in the privacy of their home, and resembling a natural miscarriage. In contrast, women who chose the surgical procedure placed more importance on an abortion method that does not involve minor side effects.

An unexpected finding is, however, that women who chose surgical compared to medical abortion more highly valued a method that allows them personal control during the abortion procedure. In a recent study of abortion providers, medical abortion was seen as a method which enhances a woman's sense of involvement and control during the abortion process (Harvey, Beckman, & Satre, 1998). Providers believed that the procedure allows women to feel in control of their own bodies. Similarly, findings from a recent qualitative study indicated that medical abortion appeals to women who want to maintain personal control (Fielding et al., 2002). In the present study the item "lets you have control during the abortion" loaded with the item "lets you know where and when the abortion is going to happen." This finding suggests that women may be interpreting the concept of control as predictability. They may perceive that medical abortion is less predictable, more out of control, than surgical abortion. Unlike medical abortion techniques, surgical abortion *does* permit a woman to "control" the time and place of the actual abortion (i.e., the removal of the conceptus) by scheduling the procedure.

It is not surprising that Factor 2, natural/avoidance of surgery, was the most important factor in terms of successful prediction. Findings from previous research consistently report that wanting to avoid surgery and wanting a procedure that is natural and not invasive are the most frequently reported reasons women give for selecting medical abortion (e.g., Beckman & Harvey, 1997; Creinin & Park, 1995; Elul et al., 2000; Fielding et al., 2002; Winikoff et al., 1998). It is noteworthy, however, that Factor 3, effectiveness and timing, did not significantly predict abortion method. A possible explanation for this finding is that certain basic characteristics of abortion procedures (e.g., effectiveness, avoidance of major health problems, used early in pregnancy) are universally important to all women seeking early pregnancy termination and, therefore, do not discriminate between the two groups of women.

Despite the finding that the underlying factors of the AAQ are useful as predictive variables for choice of abortion method, we do not necessarily believe that method attributes are the only factors that are important in the decision-making process. Previous findings indicate that a woman's perceptions and choice of an abortion method are grounded in the circumstances of her life and that multiple factors enter into the decision-making process (Harvey, Beckman, & Branch, 2002). The value that women place on specific attributes of the method reflect contextual and personal factors (e.g., residence, social support, cultural background, religion, ambivalence towards abortion, and employment). In other words, method attributes will likely have different meanings and consequences for women depending on their own personal values and life circumstances. Thus, we posit that the relative value a women places on attributes of abortion methods embodies other factors that influence method choice.

Our sample of medical abortion users was limited to women who had used methotrexate. No studies to date have compared women's experiences with methotrexate-induced abortions with those induced with mifepristone. Because the characteristics of the two types of drug-induced abortions are quite similar, findings from this study could also inform counseling and education for the provision of mifepristone-induced abortions. Research is needed, however, that examines factors that influence women's choice among different types of medical abortion as well as their choice between mifepristone abortions and surgical procedures.

A significant limitation of the present study is the ability to generalize findings to other women. First, our data were collected from women enrolled in a clinical trial. It is possible that women who are early adopters of a new technology may differ from those who will choose a method after it has become more established and familiar. Attitudes about newness and risk-taking may be important in method choice (Winikoff, 1995). Second, although the response rate was fairly high, those who agreed to participate in the study may be different from those who did not. Finally, because our sample was small and limited in diversity, our findings may not be representative of women from different racial/ethnic and other sociodemographic groups or representative of women who obtain abortions in the United States (see Jones, Darroch, & Henshaw, 2002). It is noteworthy, however, that the proportion of women in our sample who experienced a previous abortion (47%)

is consistent with national trends (48%; Jones et al., 2002). Future research on the AAQ might examine the responses of different populations to determine whether the factor structures are similar and whether the scales are valid for other groups. Additional studies might focus on ratings of abortion methods not included in the present research, particularly other methods of medical abortion.

Finally, in studies based on self-report, it is always difficult to know when participants are providing a rationale for their behavior. For example, because women completed the questionnaire after they had chosen their abortion method it could be argued that participants' responses to the importance items are justifications for their choice of method, rather than characteristics that they perceive are important in method choice. We believe, however, that the AAQ captures individuals' underlying values and these are stable over time.

Taken together, the study findings indicate that the AAQ has the potential to be an important instrument for both clinical applications and basic research. The questionnaire could be used by abortion providers to determine which characteristics are important to a woman when selecting an abortion method. This process would assist a woman in making informed decisions about abortion methods and help providers match her with a method that she finds acceptable. Findings from recent studies of abortion providers indicate that a major barrier to the provision of medical abortion is the quality and quantity of counseling needed for medical abortion patients (Breitbart & Callaway, 2000; Harvey, Beckman, & Satre, 2000; Henshaw & Finer, 2003; Joffe, 1999). Conceivably, the AAQ would reduce the time spent counseling women and provide an additional strategy to assist women in the selection of abortion procedures. We believe, however, that the AAQ should only be used as a tool to supplement, *not replace*, non-directive counseling regarding options of abortion methods. Finally, the AAQ could also be used in additional studies to obtain a valid measure of the subjective importance of characteristics of abortion methods.

References

Bachelot, A., Cludy, L., & Spira, A. (1992). Conditions for choosing between drug-induced and surgical abortions. *Contraception, 45*, 547–549.

Beckman, L. J., & Harvey, S. M. (1997). Experience and acceptability of medical abortion with mifepristone and misoprostol among U.S. women. *Women's Health Issues, 7*(4), 253–262.

Berer, M. (1992). Inducing miscarriage: Women centered perspectives on RU 486/prostaglandin as an early abortion method. *Law, Medicine and Health Care, 20*, 199–208.

Breitbart, V., & Callaway, D. (2000). The counseling component of medical abortion. *Journal of the American Medical Women's Association, 55*(3, Suppl. 2000), 164–166.

Creinin, M. D. (1993). Methotrexate for abortion at < 42 days gestation. *Contraception, 48*(4), 519–525.

Creinin, M. D., & Darney, P. D. (1993). Methotrexate and misoprostol for early abortion. *Contraception, 48*(4), 339–348.

Creinin, M. D., & Park, M. (1995). Acceptability of medical abortion with methotrexate and misoprostol. *Contraception, 52*, 41–44.

Elul, B., Pearlman, E., Sohaindo, A., Simonds, W., & Westhoff, C. (2000). In-depth interviews with medical abortion clients: Thoughts on the method and home administration of misoprostol. *Journal of the American Medical Women's Association, 55*, 169–172.

Fielding, L., Edmunds, E., & Schaff, E. A. (2002). Having an abortion using mifepristone and home misoporostol: A qualitative analysis of women's experiences. *Perspectives on Sexual and Reproductive Health, 34*(1), 34–40.

Finer, F. B., & Henshaw, S. K. (2003). Abortion incidence and services in the United States in 2000. *Perspectives on Sexual and Reproductive Health, 35*(1), 16–24.

Harvey, S. M., Beckman, L. J., & Branch, M. R. (2002). The relationship of contextual factors to women's perceptions of medical abortion. *Health Care for Women International, 23*(6–7), 654–665.

Harvey, S. M., Beckman, L. J., Castle, M. A., & Coeytaux, F. (1995). Knowledge and perceptions of medical abortion among potential users. *Family Planning Perspectives, 27*(5), 203–207.

Harvey, S. M., Beckman, L. J., & Satre, S. (1998). *Listening to and learning from health care providers about methotrexate-induced abortions* (Working Paper). Pacific Institute for Women's Health, Los Angeles.

Harvey, S. M., Beckman, L. J., & Satre, S. (2000). Experience and satisfaction with providing methotrexate-induced abortion services among U.S. providers. *Journal of the American Medical Women's Association, 55*(3, Suppl. 2000), 161–163.

Hausknecht, R. U. (1995). Methotrexate and misoprostol for abortion at 57–63 days gestation. *New England Journal of Medicine, 333*(9), 337–340.

Henshaw, S. T., & Finer, L. B. (2003). The accessibility of abortion services in the United States, 2001. *Perspectives on Sexual and Reproductive Health, 35*(1), 18–24.

Joffe, C. (1999). Reactions to medical abortion among providers of surgical abortion: An early snapshot. *Family Planning Perspectives, 31*(1), 35–38.

Jones, R. K., Darroch, J. E., & Henshaw, S. K. (2002). Patterns in the socioeconomic characteristics of women obtaining abortions in 2000–2001. *Perspectives on Sexual and Reproductive Health, 34*(5), 226–235.

Nunnally, J., & Bernstein, I. (1994). *Psychometrics Theory* (3rd ed.). New York: McGraw-Hill.

Rosen, A. S., Von Knorring, K., & Bygdeman, M. (1984). Randomized comparison of prostaglandin treatment in hospital or at home with vacuum aspiration for termination of early pregnancy. *Contraception, 29*, 423–435.

Schaff, E., Eisinger, S. H., Franks, P., & Kim, S. S. (1995). Combined methotrexate and misoprostol for early induced abortion. *Archives of Family Medicine, 4*(9), 774–779.

Tang, G. W. (1991). A pilot study of acceptability of RU 486 and ONO 802 in Chinese population.*Contraception, 44*, 523–532.

Tang, G. W., Lau, O. W. K., & Yip, P. (1993). Further acceptability evaluation of RU 486 and ONO 802 as abortifacient agents in a Chinese population. *Contraception, 48*, 267–276.

Thong, K. J., Dwar, M. H., & Baird, D. T. (1992). What do women want during medical abortion? *Contraception, 46*, 435–442.

Winikoff, B. (1995). Acceptability of medical abortion in early pregnancy. *Family Planning Perspectives, 27*, 142–148, 185.

Winikoff, B., Ellertson, C., Elul, B., & Sivin, I. (1998). Acceptability and feasibility of early pregnancy termination by mifepristone-misoprostol. *Archives of Family Medicine, 7*, 360–366.

S. MARIE HARVEY is the Director of the Research Program on Women's Health at the Center for the Study of Women in Society and Associate Professor of Public Health at the University of Oregon. She is a Fellow of the American Psychological Association (APA) and Past President the APA Division 34, Population and Environmental Psychology and Past-Chair of the Population, Family Planning and Reproductive Health Section of the American Public Health Association. Her current research interests focus on the acceptability of reproductive technologies;

the prevention of HIV/STIs among high-risk women, men, and couples; and the influence of relationship factors on sexual risk-taking. Marie currently serves as Principal Investigator on a five-year project that assesses predictors of sexual risk behavior and designs, implements and evaluates a couple-based intervention to reduce unprotected intercourse. She is also Principal Investigator on "Women's Acceptability of the Vaginal Diaphragm," a three year study recently funded by the National Institute for Child Health and Human Development.

MARK NICHOLS, is Associate Professor on the faculty of the Ob/Gyn Department of Oregon Health Sciences University and the chief of their division of General Gynecology and Obstetrics. He also is the Medical Director of Planned Parenthood of the Columbia Willamette. He was a member of the National Medical Committee of the Planned Parenthood Federation of America from 1996 to 2002, serving as chair of the nominating committee. His research interests are in teaching medical students and residents family planning and abortion, reduction of pain during surgical abortion, and hormonal contraception.

Appendix. Items on the Abortion Attribute Questionnaire (AAQ)

1.	*The method* is very effective.
2.	*The method* can be used very early in pregnancy.
3.	*The method* lets you have the abortion over with quickly.
4.	*The method* doesn't have side effects like nausea, headache, and diarrhea.
5.	*The method* doesn't involve heavy bleeding.
6.	*The method* hardly ever causes major health problems (infection, tearing the uterus).
7.	*The method* doesn't involve surgery.
8.	*The method* lets you have an abortion without people you are close to finding out about it.
9.	*The method* lets you have the abortion in the privacy of your home.
10.	*The method* can be used later in pregnancy (after 9 weeks).
11.	*The method* doesn't involve seeing the blood and blood clots during the abortion.
12.	*The method* doesn't involve painful cramping for more than an hour.
13.	*The method* doesn't involve bleeding longer than 7 days.
14.	*The method* has been used for abortion for a long time.
15.	*The method* takes only one or two office visits.
16.	*The method* has a doctor or nurse around during the abortion.
17.	*The method* lets you have control during the abortion.
18.	*The method* doesn't involve surgical instruments being inside you.
19.	*The method* lets you know where and when the abortion is going to happen.
20.	*The method* is like a natural miscarriage.
21.	*The method* is inexpensive.

Journal of Social Issues, Vol. 61, No. 1, 2005, pp. 109–126

Conspiracy Beliefs About HIV/AIDS and Birth Control Among African Americans: Implications for the Prevention of HIV, Other STIs, and Unintended Pregnancy

Sheryl Thorburn Bird*
Oregon State University

Laura M. Bogart
RAND Corporation

In this article, we examine the potential role that conspiracy beliefs regarding HIV/AIDS (e.g., "HIV is a manmade virus") and birth control (e.g., "The govern-ment is trying to limit the Black population by encouraging the use of condoms") play in the prevention of HIV, other STIs, and unintended pregnancies among African Americans in the United States. First, we review prior research indicat-ing that substantial percentages of African Americans endorse conspiracy beliefs about HIV/AIDS and birth control. Next, we present a theoretical framework that suggests how conspiracy beliefs influence sexual behavior and attitudes. We then offer several recommendations for future research. Finally, we discuss the policy and programmatic implications of conspiracy beliefs for the prevention of HIV, other STIs, and unintended pregnancy.

African Americans are disproportionately affected by HIV/AIDS, other sex-ually transmitted infections (STIs), and unintended pregnancy. More specifically, HIV and AIDS rates are substantially higher among African Americans than among Hispanics or Whites in the United States (Centers for Disease Control and Prevention [CDC], 2001). Although about 12% of the total U.S. population is

---

*Correspondence concerning this article should be addressed to Sheryl Thorburn Bird, Department of Public Health, Oregon State University, 264 Waldo Hall, Corvallis, Oregon 97331-6406 [e-mail: Sheryl.Thorburn@oregonstate.edu].

non-Hispanic Black (U.S. Census Bureau, 2001), non-Hispanic Blacks account for approximately 49% of U.S. AIDS cases and 50% of U.S. HIV infection cases reported in 2001 (CDC, 2001). STI rates also are generally higher among African Americans, in part, but not entirely, due to differences in reporting (CDC, 2000). For example, rates of gonorrhea and syphilis are 30 times higher among African Americans than among Whites (CDC, 2000). Although many STIs are initially asymptomatic, they can have serious health consequences (e.g., cancer, infertility) years later, particularly for women (Eng & Butler, 1997). Moreover, untreated STIs can facilitate HIV transmission by increasing HIV infectivity and susceptibility (e.g., Fleming & Wasserheit, 1999; Rothenberg, Wasserheit, St. Louis, Douglas, & the Ad Hoc STI/HIV Transmission Group, 2000). Furthermore, in the United States, the unintended pregnancy rate among Black women is estimated to be over twice the rate among White women (Henshaw, 1998).

Given the large racial/ethnic disparities in HIV, other STIs, and unintended pregnancies, it is essential to identify the factors that contribute to high-risk sexual behavior among African Americans. Prior research has described conspiracy beliefs held by some African Americans regarding HIV/AIDS (e.g., "HIV is a manmade virus") and birth control (e.g., "The government is trying to limit the Black population by encouraging the use of condoms") that may be barriers to HIV, STI, and pregnancy prevention (e.g., Herek & Capitanio, 1994; Klonoff & Landrine, 1999a; Thomas & Quinn, 1993). Conspiracy beliefs are beliefs about large-scale discrimination, by the government and health care system, against a group (in this case, African Americans). Such beliefs have important implications for policy and prevention programs.

In this article, we examine the potential role that conspiracy beliefs play in the prevention of HIV, other STIs, and unintended pregnancy among African Americans. Although researchers have documented HIV/AIDS conspiracy beliefs in Africa (e.g., Castle, 2003), our discussion focuses on beliefs held by African Americans in the United States. We begin by reviewing what is currently known about the prevalence of conspiracy beliefs regarding HIV/AIDS and birth control and their relationship to sexual behavior, including some findings from our recent research on this topic. We then discuss a theoretical framework, which draws from social psychological theories regarding social cognitive expectancies. Next, we offer several recommendations for future research. Finally, we discuss the policy and programmatic implications of conspiracy beliefs for the prevention of HIV, other STIs, and unintended pregnancy.

Research on Conspiracy Beliefs about HIV/AIDS

Much commentary on HIV prevention has emphasized the need to take into account conspiracy beliefs in the African American community when designing culturally-relevant interventions to increase safer sexual behavior (e.g., Clark,

1998; McGary, 1999; Smith, 1999; Thomas & Quinn, 1991). However, almost no research has examined the relationship between HIV/AIDS conspiracy beliefs and sexual behavior; instead, the focus has been on documenting the prevalence of such beliefs in the African American community.

Several surveys have noted that a significant percentage of African Americans hold conspiracy beliefs regarding HIV/AIDS, and that such beliefs are more prevalent among Blacks than among Whites (e.g., Crocker, Luhtanen, Broadnax, & Blaine, 1999). Herek and Glunt (1991) conducted a telephone survey with a random sample of U.S. households in which they found that two-thirds of Blacks (67%) compared to one-third of Whites (34%) agreed that the government is not telling the whole story about AIDS. Herek and Capitanio (1994) found in a telephone survey of Blacks and Whites in the United States that 20% of Blacks agreed with the statement "The government is using AIDS as a way of killing off minority groups" compared to 4% of Whites. In addition, 43% of African Americans and 37% of Whites agreed that "A lot of information about AIDS is being held back from the general public." Lower levels of education and income were associated with the belief that the government is using AIDS as a way of killing off minority groups. Neither of the conspiracy beliefs was significantly related to self-reported behavior change as a result of AIDS (i.e., changes in respondents' risk behavior as a result of AIDS, such as reducing their numbers of sexual partners or increasing their use of condoms).

Thomas and Quinn (1993) reported findings regarding HIV/AIDS conspiracy beliefs from a number of surveys with diverse Black samples in the United States (e.g., community samples, random sample of Black households). Depending on the sample, between 28% and 44% of respondents indicated that they did not trust government reports about AIDS. Between 17% and 38% of respondents believed that there is some truth in reports that the AIDS virus was produced in a germ warfare laboratory. Between 15% and 35% agreed that AIDS is a form of genocide against the Black race. Similarly, Parsons, Simmons, Shinhoster, and Kilburn (1999) surveyed parishioners of 35 churches in Louisiana. Their analyses focused on data from respondents who were African American adults (excluding visitors from other states). They examined respondents' endorsement of a wide range of conspiracy beliefs and found that almost 70% of respondents did not believe that the government is telling the truth about AIDS, and over 25% agreed that AIDS was "intended to wipe Blacks off the face of the earth."

Klonoff and Landrine (1999a) surveyed Black adults door-to-door in 10 randomly selected middle- and working-class census tracts in San Bernardino County, California. In self-administered questionnaires, 14% of respondents totally agreed, and 12% somewhat agreed, with the statement "HIV/AIDS is a man-made virus that the federal government made to kill and wipe out Black people." The statement was most likely to be endorsed by individuals who were culturally traditional (i.e., immersed in Black culture), men, individuals who more frequently perceived discrimination in their lifetime, and individuals who had higher levels of education (i.e., college versus high school graduates). Among men, reports of more frequent perceived lifetime discrimination and less frequent perceived recent discrimination were associated with agreeing with the conspiracy belief. Among women, being culturally traditional was associated with endorsing the conspiracy belief. Klonoff and Landrine's findings for education were inconsistent with those of Herek and Capitanio (1994), which could be due to sample differences, temporal effects, or the influence of unmeasured variables. Klonoff and Landrine suggested that individuals with more education might be especially aware of how others treat them in society and may have greater knowledge of government-based conspiracy theories. However, additional research, which examines a range of conspiracy beliefs, is clearly needed before conclusions can be made about the nature of the relationship between education level or other measures of socioeconomic status and belief in HIV/AIDS conspiracies.

In sum, prior work has begun to document the prevalence of HIV/AIDS conspiracy beliefs among African Americans. Although these types of beliefs have obvious implications for HIV prevention, we found only one study that examined the relationship between belief in an HIV/AIDS conspiracy and sexual behavior, and the association was not significant (Herek & Capitanio, 1994). However, the researchers did not examine a full range of sexual behaviors and attitudes, such as attitudes toward condom use and condom use behaviors. Such research would be useful in pinpointing the influence of HIV/AIDS conspiracy beliefs on sexual health and sexual behavior related to HIV.

Another understudied area in this realm is the implications of conspiracy beliefs for the treatment behaviors of people with HIV. Several conspiracy beliefs about antiretroviral medications for HIV have been noted in the popular press, including the belief that the drugs can kill patients (France, 1998), and that people who take the medications are human guinea pigs for the U.S. government (Richardson, 1997). Almost no research has documented conspiracy beliefs regarding these medications in a systematic fashion, or examined whether such beliefs are related to treatment behaviors, among those who are infected with HIV. Given that African Americans are less likely than Whites in the United States to adhere to HIV treatment and to participate in clinical drug trials for HIV medications (e.g., Bogart, Bird, Walt, Delahanty, & Figler, 2004; Gifford et al., 2002; Singh et al., 1996; Stone, Mauch, Steger, Janas, & Craven, 1997), research is needed to assess whether conspiracy beliefs about HIV treatment contribute to racial/ethnic differences in HIV treatment behavior.

To extend previous work, we recently conducted a telephone survey to explore the relationship of HIV/AIDS and birth control conspiracy beliefs to sexual behavior and attitudes among a national, randomly selected sample of African Americans (aged 18–45 years; $N = 71$) in the United States (Bird & Bogart, 2003; Bogart & Bird, 2003). Our goal was to inform the design of a larger-scale study on this topic. We examined two types of HIV/AIDS conspiracy beliefs: beliefs about the governments' role in the AIDS epidemic (e.g., AIDS is a form of genocide against African Americans) and beliefs about the efficacy of new treatments for HIV (e.g., "The medicine that doctors prescribe to treat HIV is poison."). In general, we found strong endorsement of many of the beliefs. For example, 70% of respondents somewhat or strongly believed that "A lot of information about AIDS is being held back from the public," and about half (53%) somewhat or strongly endorsed the statement that "There is a cure for AIDS, but it is being withheld from the poor." Although fewer participants endorsed medication-related conspiracy beliefs, over 40% somewhat or strongly agreed that "People who take the new medicines for HIV are human guinea pigs for the government."

We found somewhat different patterns of results for the two types of HIV/AIDS conspiracy beliefs. Belief in HIV government conspiracies was related to less positive attitudes toward condoms for birth control and greater numbers of partners in the past three months, suggesting that mistrust of government institutions reduces openness to public health prevention messages regarding HIV. In contrast, belief in HIV/AIDS treatment conspiracies was related to more positive attitudes about using condoms in the next three months and a greater reported likelihood of using condoms at next intercourse among all respondents. We speculate that individuals who do not trust new treatments for HIV may be motivated to use condoms in order to avoid those treatments. Even though our study was cross-sectional and had a small sample, our findings indicate that further work in this area is warranted.

In addition to conspiracy beliefs about HIV/AIDS, a small amount of research has documented the existence of conspiracy beliefs about birth control among African Americans. In the following section, we discuss the etiology and content of these beliefs, as well as the limited amount of research on this topic.

Research on Conspiracy Beliefs about Birth Control

It should not be surprising that some African Americans have concerns about contraceptive methods and the underlying motives of pregnancy prevention programs. First of all, dating back to slavery, the United States has a long history of racist and discriminatory behavior with respect to efforts to control African American women's fertility (Roberts, 1998). More recently, in the 1900s, as part of the eugenics and social control movements and later in efforts to link contraception to welfare, African American women and poor women were disproportionately the target of programs and legislative proposals in the United States promoting, coercing, or requiring contraceptive use, particularly methods such as female sterilization and Norplant (Boonstra et al., 2000; Brown & Eisenberg, 1995; Dugger, 1998; Roberts, 1998, 2000; Rodrigues-Triaz, 1982). Birth control pills and Norplant were considered by some to be important new technological developments that could help combat overpopulation and poverty, and some policy-makers incorrectly associated such social problems exclusively with African Americans (Boonstra et al., 2000; Gamble, 1994; Roberts, 1998).

The belief that birth control and family planning programs were part of a genocidal conspiracy against African Americans was discussed in the public health and medical literature in the 1960s and 1970s (e.g., Allen, 1977; Lincoln, 1975; Treadwell, 1972; Weisbord, 1973). More recently, Turner (1993) described how, after Norplant's legalization in 1990, rumors that African American and other minority women in the United States were being forced to have Norplant implanted began to circulate. Turner reports that a version of this rumor linked Norplant to Black genocide. She collected these rumors almost a year after policy-makers and others made statements recommending that women on welfare be required or given financial incentives to use Norplant.

Even though scholars have previously described beliefs linking contraception and family planning to Black genocide (Brown & Eisenberg, 1995; Dugger, 1998; Roberts, 1998, 2000; Turner, 1993; Weisbord, 1973), research investigating conspiracy beliefs about birth control is limited. In the early 1970s, Darity and Turner (1972) reported on a survey conducted with a sample of households from Black neighborhoods, stratified according to low-income versus middle-to-upper income, in a medium-sized New England city. Interviews were conducted with the head of household or a female member of the household who was of reproductive age. They examined agreement with seven statements regarding racial genocide. Four of the statements were conspiracy beliefs regarding birth control and abortion: "Encouraging American blacks to use birth control is comparable to trying to eliminate this group from society," "Abortions are a part of a white plot to eliminate blacks," "Sterilization is a white plot to eliminate blacks," and "All forms of birth control methods are designed to eliminate black Americans." They found that each of these statements was associated with reports of not using family planning (birth control) methods.

A year later, Turner and Darity (1973) reported results from a larger study of Blacks living in Philadelphia, Pennsylvania, and Charlotte, North Carolina. The goal of the study was to assess the extent of genocide fears among Black Americans and explore the relationships of sex, age, and region to such fears. Households were randomly selected from two low socioeconomic status (SES) communities and two middle-to-high SES communities which had populations that were at least 50% Black. In each household, a woman 15–44 years of age and, in most cases, her most significant male partner were interviewed. It is important note that a limitation of this study is that the men and women in the sample were not independent. Even so, the study's results indicated that large percentages of respondents endorsed the following conspiracy beliefs: "Birth control clinics in black neighborhoods should be operated by Blacks" (women, 54.7%; men 62.3%); "As the need for cheap labor goes down, there will be an effort to reduce the number of blacks" (women, 54.1%;

men 47.7%); "As blacks become more militant, there will be an effort to decrease the black population" (women, 53.4%; men 61.7%); "The survival of black people depends on increasing the number of black births" (women, 50.8%; men 56.1%); and "Birth control programs are a plot to eliminate blacks" (women, 37.2%; men 41.3%). In addition, controlling for gender, Turner and Darity found some age and regional differences in level of agreement.

Similarly, Farrell and Dawkins (1979) conducted a survey with a stratified sample of Black households in a county in southeastern Texas. In each household, a woman 15–44 years of age and, in most cases, her most significant male partner were interviewed. Farrell and Dawkins found that a significantly greater percentage of non-users than users of birth control agreed that birth control is a plot against Blacks (56% vs. 23%), that Black survival depends on increasing the numbers of Blacks (64% vs. 47%), that abortion is a plot against Blacks (59% vs. 24%), and that birth control clinics should be run by Blacks (77% vs. 61%). They did not, however, examine the relationship between these beliefs and birth control use separately for the men and women in the sample. Because the men and women interviewed were not independent, these results should be viewed with caution.

More recently, in their 1996 survey of church parishioners in Louisiana, Parsons et al. (1999) found that over 25% of respondents agreed that family planning programs were a form of genocide. Their sample was limited to church parishioners in a single state, and they examined only one conspiracy belief regarding birth control. However, the results from this study suggest that birth control conspiracy beliefs are currently shared by at least some African Americans, and that such beliefs deserve further attention.

In summary, prior research suggests that some African Americans endorse conspiracy beliefs regarding birth control. However, a small number of studies have examined conspiracy beliefs about birth control, and only one was conducted within the last 20 years. In addition, each of these studies has limitations, especially with respect to sample design and the lack of analyses examining the relationship between birth control conspiracy beliefs and sexual behavior and related outcomes.

As noted earlier, in order to extend previous research, we recently conducted an exploratory study that examined birth control conspiracy beliefs among a small national sample of African Americans (Bird & Bogart, 2003). We assessed the extent to which respondents agreed with six statements capturing conspiracy beliefs related to birth control. We found that many of the respondents endorsed the beliefs. More specifically, 37% somewhat or strongly agreed that "Having children is the key to the survival of the African American population" and 49% somewhat or strongly agreed that "Whites want to keep the numbers of African American people down." In addition, 27% somewhat or strongly endorsed the belief that "Poor and minority women are sometimes forced to be sterilized by the government," 21% somewhat or strongly agreed that "The government is trying to limit the African American population by encouraging the use of condoms," and 35% somewhat or strongly agreed that "Medical and public health institutions use poor and minority people as guinea pigs to try out new birth control methods." Only 6%, however, endorsed the belief that "Birth control is a form of Black genocide."

We also found that birth control conspiracy beliefs were associated with contraceptive behavior and attitudes. In general, stronger endorsement of birth control conspiracy beliefs was associated with more negative attitudes toward contraceptive methods, including birth control pills and condoms. Furthermore, our findings supported the hypothesis that birth control conspiracy beliefs influence contraceptive choices. For example, we found that more strongly agreeing with the statement "Medical and public health institutions use poor and minority people as guinea pigs to try out new birth control methods" was associated with currently using condoms for contraception among birth control users. A possible explanation for this finding is that African Americans with stronger birth control conspiracy beliefs may be more suspicious of certain contraceptive methods and, if they want to use birth control, may be more likely to choose a method that does not require a visit to a health care provider and/or is not a hormonal method (e.g., condoms). In addition, our results indicated that conspiracy beliefs about birth control are related to intentions to use specific contraceptive methods. For example, greater endorsement of the statement "Birth control is a form of Black genocide" was associated with lower intentions to use birth control pills in the next year. Further research is needed in order to determine the extent to which conspiracy beliefs about birth control are endorsed in the United States today, and to better understand the association between these beliefs and contraceptive attitudes and behavior, as well as utilization of reproductive health services. Such relationships would have implications for pregnancy prevention programs, the introduction of new contraceptive methods, and the delivery of contraceptive services.

Theoretical Framework

A number of researchers have suggested that racial/ethnic differences in health care-related attitudes and behaviors are a direct consequence of decades of discrimination in the United States by Whites in general, the U.S. government, and the health care system in particular (Krieger, 2000; Landrine & Klonoff, 2001). Research shows that a substantial percentage of African Americans have experienced some form of discrimination in their lives, in the recent past, and within the health care system (e.g., Bird & Bogart, 2001; Klonoff & Landrine, 1999b; Krieger & Sidney, 1996; Landrine & Klonoff, 1996). For example, Landrine and Klonoff (1996) surveyed African American students, faculty, and staff at a large university in the United States and found that all respondents reported experiencing racist discrimination in their lifetime, and 98% reported experiencing racist discrimination in the past year. Furthermore, in a study with an African American community sample in northeast Ohio, we found that almost two-thirds of participants reported having experienced discrimination within the health care context because of their race (Bird & Bogart, 2001).

Prior research suggests that discrimination by health care providers plays a significant role in African Americans' negative health outcomes (e.g., Bogart, Catz, Kelly, & Benotsch, 2001; van Ryn & Burke, 2000). For example, in a recent study conducted with physicians and patients in New York State hospitals, physicians tended to rate their African American patients more negatively than their White patients on several dimensions, including but not limited to intelligence, education, and likelihood of risk behavior and compliance with medical advice, even after controlling for a variety of patient and physician characteristics (van Ryn & Burke, 2000). Moreover, other objective indicators of health care disparities in the United States show that African Americans receive worse medical treatment, use fewer medical services, and have worse health outcomes than do Whites, even after controlling for current health and socioeconomic status (for reviews of the literature, see Mayberry, Mili, & Ofili, 2000; Smedley, Stith, & Nelson, 2002).

In addition to personal, first-hand experiences with discrimination, African Americans' awareness of institutional health care discrimination has also contributed to their suspicions about the U.S. government and health care system. In particular, knowledge of the Tuskegee Syphilis Study is a significant source of mistrust of U.S. medical and public health institutions among African Americans (e.g., Gamble, 1997; Jones, 1993; Landrine & Klonoff, 2001; Thomas & Quinn, 1991). In the Tuskegee Syphilis Study, the U.S. Public Health Service investigated the effects of untreated syphilis in African American men for 40 years, from 1932 to 1972. Despite the availability of a treatment for syphilis after 1947, participants were denied treatment and merely monitored throughout the course of their disease. The researchers continued to observe participants as they went blind, insane, and eventually died.

Given the historical and present-day context of African Americans' experiences with discrimination, it is not surprising that mistrust of Whites and suspicion of White institutions, such as the U.S. health care system, are prevalent in African American culture (Landrine & Klonoff, 2001). A significant percentage of African Americans endorse beliefs indicative of institutional racism (i.e., beliefs about the discriminatory policies carried out by state or non-state organizations; Krieger, 2000). For example, African Americans perceive more racism in health care and report greater mistrust of the medical system than do Whites (Finnegan et al., 2000; LaVeist, Nickerson, & Bowie, 2000; Lillie-Blanton, Brodie, Rowland, Altman, & McIntosh, 2000). It is in this context that conspiracies about the U.S. health care system have arisen.

How might conspiracy beliefs influence behaviors and attitudes related to the prevention of HIV, other STIs, and unintended pregnancy? Social psychological models of health behavior suggest that social cognitive expectancies about one's illness and the health care process play a role in health behaviors and, ultimately, health outcomes (e.g., Ajzen, 1985; Ajzen & Fishbein, 1980; Bandura, 1977a, 1977b, 1986, 1992; Janz & Becker, 1984; Safer, Tharps, Jackson, & Leventhal, 1979). An expectancy is "any belief, hypothesis, theory, assumption, or accessible construct that is brought from previous experience and used, either consciously or unconsciously, as a basis for interpreting or generating behavior in the present context" (Ditto & Hilton, 1990, p. 99). As Ditto and Hilton (1990) describe, expectancies can be impersonal such as beliefs about a specific disease or about particular institutions. Conspiracy beliefs about HIV/AIDS and birth control are impersonal expectancies that may play a major role in HIV, STI, and pregnancy prevention behavior and attitudes.

More specifically, conspiracy beliefs about the government's role in HIV/AIDS may be a barrier to HIV prevention because, in the United States, medical and public health institutions are a major source of information about, and have a major role in the prevention and treatment of, HIV/AIDS. For example, someone who endorses HIV/AIDS conspiracy beliefs may be less open to HIV prevention messages endorsed by the U.S. government; that is, they may view condoms more negatively and may be less likely to practice safer sexual behaviors. Similarly, conspiracy beliefs about the government's role in birth control could be a barrier to the prevention of unintended pregnancy. More specifically, individuals who endorse birth control conspiracy beliefs may be less receptive to pregnancy prevention. They may have more negative attitudes toward contraception and may avoid visiting health care providers for contraceptive services. In this way, conspiracy beliefs are likely to affect intentions to engage in safer sexual behavior and to seek reproductive health services.

Recommendations for Research

Taken together, findings from previous research suggest that conspiracy beliefs about HIV/AIDS and birth control may play an important role in African Americans' sexual risk behavior. Reports of conspiracy beliefs in the media and anecdotally, and the concerns that such beliefs raise among the public as well as public health professionals, heighten the significance of research in this area. In this section, we present several recommendations for future research.

First, more research is needed to determine the prevalence of conspiracy beliefs regarding HIV/AIDS and birth control in the United States. Although the study on AIDS conspiracy beliefs by Herek and Capitanio (1994) collected data from a national sample, it was conducted over 10 years ago. Similarly, the surveys reported on by Thomas and Quinn (1993) occurred more than a decade ago, and the samples for those surveys and the studies by Parsons and associates (1999) and Klonoff and Landrine (1999a) were not representative of the U.S. population. The existing research on birth control conspiracy beliefs has similar shortcomings, and some of those studies have additional design limitations (Darity & Turner, 1972; Farrell & Dawkins, 1979; Parsons et al., 1999; Turner & Darity, 1973). In the past 20 years, scientific knowledge about HIV/AIDS has dramatically increased. Furthermore, there have been major advances in the diversity and availability of contraceptive methods since the 1970s, such as the introduction of Norplant and Depo Provera (Boonstra et al., 2000; Piccinino & Mosher, 1998). These changes are likely to have influenced the prevalence and nature of conspiracy beliefs regarding HIV/AIDS and birth control during the past 20 years. Thus, research is needed that determines the degree to which conspiracy beliefs are endorsed by the U.S. population, including—but not limited to—African Americans, and the nature of those beliefs. Studies that focus on high-risk populations, other racial/ethnic groups (e.g., Hispanics), and other stigmatized groups (e.g., women on Medicaid) should be conducted also. In addition, determining whether conspiracy beliefs contribute to racial/ethnic differences in sexual behavior, attitudes, and related reproductive health behavior is important.

Further research is needed, also, in order to understand how conspiracy beliefs affect high-risk sexual behavior. In particular, prospective studies that examine the relationship between conspiracy beliefs and later sexual behavior are essential. For example, until there are longitudinal data, we will not know whether conspiracy beliefs about birth control influence contraceptive behavior, whether experiences obtaining or using particular contraceptive methods affect individuals' beliefs about birth control conspiracies, or both.

Studies that identify the sources of conspiracy beliefs and examine their diffusion in African American communities could help inform efforts to prevent the emergence and spread of new conspiracy beliefs in the future. For example, increased understanding of how conspiracy theories regarding condoms and other contraceptive methods are initiated could be useful in planning the introduction and promotion of new HIV prevention methods such as microbicides (i.e., chemical barrier methods). The extent to which the behavior of public health officials, health care providers, policy-makers, political leaders, and governmental agencies contributes to and fuels conspiracy theories must also be examined. Future research should also examine the relationships among conspiracy beliefs, perceived discrimination, mistrust of medical and public health institutions, and sexual risk behavior. Because conspiracy beliefs are hypothesized to stem from chronic experiences of discrimination, it would be useful to test a prospective model that examines the dynamic relationships among perceived discrimination, mistrust of health care, and belief in conspiracies and the effects of these types of beliefs on sexual behavior. Finally, the effectiveness of different strategies to dispel conspiracy beliefs should be evaluated.

Even though significant gaps in knowledge about HIV/AIDS and birth control conspiracy beliefs exist, past research suggests that these beliefs are prevalent. As discussed in the following section, these beliefs must be addressed in prevention efforts.

Policy and Programmatic Implications

Conspiracy beliefs about HIV/AIDS and birth control have important implications for policy and programs directed at the prevention of HIV, other STIs, and unintended pregnancy among African Americans. In this section, we discuss some of those implications.

Culturally-Sensitive Interventions for Increasing Condom Use and Reducing High-Risk Sexual Behaviors

Given the current and historical mistreatment of African Americans by the health care system, it is understandable that people believe in conspiracies regarding HIV/AIDS and birth control. Such conspiracy beliefs should not be ignored. Because research indicates that substantial proportions of African Americans share conspiracy beliefs, culturally-sensitive interventions and educational campaigns should address these beliefs. More specifically, interventions and educational programs directed at HIV and STI prevention need to provide accepting environments in which to discuss conspiracy beliefs and sources of mistrust of medical and public health institutions and of Whites (e.g., the Tuskegee Syphilis Study) in combination with the facts about HIV such as the biological pathways of transmission. Similarly, pregnancy prevention programs should address birth control conspiracy beliefs as well as discuss the characteristics of different contraceptives.

Because of the high levels of mistrust in African American communities regarding public health messages advanced by the U.S. government, prevention interventions should be delivered by individuals within those communities, rather than unknown "outsiders" who are part of the larger public health system. Community-based interventions have been shown to be successful in reducing risky behaviors in a number of at-risk populations, including inner-city women. In these types of interventions, prevention education is delivered by peers who are agreed-upon popular opinion leaders in their community (CDC, 1996; Kelly et al., 1992, 1997; Sikkema et al., 2000). For example, in a study of inner-city women in five U.S. cities, a cadre of popular opinion leaders were recruited from their housing projects to organize neighborhood events, such as picnics and presentations, that educated members of the community about HIV (Sikkema et al., 2000). In comparison to women in housing projects who had not received the popular opinion leader intervention, women who received the intervention were less likely to have engaged in any unprotected intercourse and had a higher percentage of protected acts of intercourse at follow-up.

Thus, community-based interventions have demonstrated efficacy at reducing high-risk sexual behaviors and may be an ideal context in which to address conspiracy beliefs. If conspiracy beliefs are addressed in a neighborhood context by respected peers, there may be greater acceptance of the prevention message. Peer educators may be seen as more credible than members of the public health system when advancing HIV, STI, and pregnancy prevention messages.

Introducing and Promoting Methods for Preventing HIV and Unintended Pregnancy

Government and industry need to anticipate and address conspiracy beliefs when planning for the introduction of new contraceptives and HIV prevention methods. For example, in our research, we found that a substantial proportion (35%) of African Americans believed that medical and public health institutions use poor and minority women as guinea pigs to test new birth control methods (Bird & Bogart, 2003). This finding suggests that the public may need more information than has previously been provided to potential consumers about the safety and effectiveness of new contraceptive methods and about the research upon which such information is based. Accordingly, when microbicides and other new HIV prevention methods are approved and become available in the United States, media campaigns and educational materials about those products will need to address potential concerns about how and upon whom they were tested, as well as to provide data on their safety and effectiveness.

Furthermore, policies and programs intended to increase use of birth control and prevent pregnancy should not conceptually link the use of contraception to the reduction of poverty or other social problems. In addition, they should not use incentives or in any way coerce women to use contraception or specific contraceptive methods. Such strategies are misguided and are, understandably, a source of mistrust of the government, the health care system, and other pregnancy prevention programs. Furthermore, programs and policies that coerce or penalize women based on their contraceptive behavior have the potential to adversely impact legitimate efforts to make contraceptives available to women and men who want them. Instead, programs and policies should increase access to safe contraceptive methods, and should be driven by concern about reproductive health rather than population control.

Delivery of Contraceptive Services

Men's and women's experiences when seeking contraceptive services may provide support for birth control conspiracy beliefs. For example, researchers have described White stereotypes about the sexuality of Black women (e.g., the stereotype that they are sexually aggressive) as well as stereotypes about welfare mothers (e.g., the stereotype that welfare mothers are African American women who have children in order to get federal funds; Gamble, 1997; Roberts, 1998; St. Jean & Feagin, 1998; Taylor, 1999). As a result of such stereotypes, health care providers and staff may make incorrect assumptions about an African American woman's reproductive health needs or background (e.g., the number of children she has, her health insurance). These assumptions may lead to or reinforce a patient's negative beliefs about health care providers' motivations and treatment of African Americans. Provider training and education could dispel stereotypes about African American and poor women, increase awareness about conspiracy beliefs, and increase providers' knowledge of the sources of African Americans' mistrust. Such training could improve patient-provider relations and providers' understanding of the contraceptive attitudes and behavior of African Americans. In a similar vein, women should be told about all of their contraceptive options, so that they can make informed decisions when selecting a contraceptive method. Although providers have a very important role in helping women choose a contraceptive method, it is equally important that providers' preferences for certain methods, or their beliefs about what methods are best for particular "types" of women, do not limit women's options.

Conclusion

Reducing unprotected intercourse and other high-risk sexual behaviors among men and women at risk of HIV, other STIs, and unintended pregnancy is a public health priority. Conspiracy beliefs regarding HIV/AIDS and birth control may play an important role in the sexual risk behavior of African Americans, by decreasing African Americans' openness to public health prevention messages. Researchers, policy-makers, and practitioners need to consider the potential effects of conspiracy beliefs and related factors on attitudes toward and the use of HIV and pregnancy prevention methods. Given the high prevalence of such beliefs among African Americans, we believe that culturally-tailored prevention interventions that are delivered by respected members of the community and incorporate frank discussion of conspiracy beliefs have the highest probability of achieving success.

References

Ajzen, I. (1985). From intentions to actions: A theory of planned behavior. In J. Kuhl & J. Beckman (Eds.), *Action-control: From cognition to behavior* (pp. 11–31). Heidelberg, Germany: Springer.
Ajzen, I., & Fishbein, M. (1980). *Understanding attitudes and predicting social behavior*. Englewood Cliffs, NJ: Prentice Hall.
Allen, J. E. (1977). An appearance of genocide: A review of governmental family-planning program policies. *Perspectives in Biology and Medicine, 29*, 300–306.
Bandura, A. (1977a). Self-efficacy: Toward a unifying theory of behavioral change. *Psychological Review, 84*, 191–215.

Bandura, A. (1977b). *Social learning theory.* Englewood Cliffs, NJ: Prentice Hall.

Bandura, A. (1986). *Social foundations of thought and action: A social cognitive theory.* Englewood Cliffs, NJ: Prentice Hall.

Bandura, A. (1992). A social cognitive approach to the exercise of control over AIDS infection. In R. J. DiClemente (Ed.), *Adolescents and AIDS: A generation in jeopardy* (pp. 89–116). London: Sage.

Bird, S. T., & Bogart, L. M. (2001). Perceived race-based and socioeconomic status (SES)-based discrimination in interactions with health care providers. *Ethnicity & Disease, 11,* 554–563.

Bird, S. T., & Bogart, L. M. (2003). Birth control conspiracy beliefs, perceived discrimination, and contraception among African Americans: An exploratory study. *Journal of Health Psychology, 8,* 263–276.

Bogart, L. M., & Bird, S. T. (2003). Exploring the relationship of conspiracy beliefs about HIV/AIDS to sexual behaviors and attitudes among African American adults. *Journal of the National Medical Association, 95,* 1057–1065.

Bogart, L. M., Bird, S. T., Walt, L. C., Delahanty, D. L., & Figler, J. (2004). Association of stereotypes about physicians to health care satisfaction, help-seeking behavior, and adherence to treatment. *Social Science & Medicine, 58,* 1049–1058.

Bogart, L. M., Catz, S. L., Kelly, J. A., & Benotsch, E. G. (2001). Factors influencing physicians' judgments of adherence and treatment decisions for patients with HIV disease. *Medical Decision Making, 21,* 28–36.

Boonstra, H., Duran, V., Gamble, V. N., Blumenthal, P., Dominguez, L., & Pies, C. (2000). The "boom and bust phenomenon": The hopes, dreams, and broken promises of the contraceptive revolution. *Contraception, 61,* 9–25.

Brown, S. S., & Eisenberg, L. (1995). *The best intentions: Unintended pregnancy and the well-being of children and families.* Washington, DC: National Academy Press.

Castle, S. (2003). Doubting the existence of AIDS: A barrier to voluntary HIV testing and counselling in urban Mali. *Health Policy and Planning, 18,* 146–155.

Centers for Disease Control and Prevention. (1996). Community-level prevention of human immunodeficiency virus infection among high-risk populations: The AIDS community demonstration projects. *Morbidity and Mortality Weekly Report, 45,* 1–24.

Centers for Disease Control and Prevention. (2000). *Tracking the hidden epidemics:Trends in STDs in the United States 2000.* Retrieved December 1, 2004, from http://www.cdc.gov/nchstp/dstd/Stats_Trends/Trends2000.pdf

Centers for Disease Control and Prevention. (2001). *HIV/AIDS Surveillance Report, 13*(2), 1–44.

Clark, P. A. (1998). A legacy of mistrust: African-Americans, the medical profession, and AIDS. *Linacre Quarterly, 87,* 66–88.

Crocker, J., Luhtanen, R., Broadnax, S., & Blaine, B. E. (1999). Belief in U.S. government conspiracies against Blacks among Black and White college students: Powerlessness or system blame? *Personality and Social Psychology Bulletin, 25,* 941–953.

Darity, W. A., & Turner, C. B. (1972). Family planning, race consciousness and the fear of race genocide. *American Journal of Public Health, 62,* 1454–1459.

Ditto, P. H., & Hilton, J. L. (1990). Expectancy processes in the health care interaction sequence. *Journal of Social Issues, 46,* 97–124.

Dugger, K. (1998). Black women and the question of abortion. In L. J. Beckman & S. M. Harvey (Eds.), *The new civil war: The psychology, culture and politics of abortion* (pp. 107–131). Washington, DC: American Psychological Association.

Eng, T., & Butler, W. T. (1997). *The hidden epidemic: Confronting sexually transmitted diseases.* Washington, DC: Institute of Medicine, National Academy Press.

Farrell, W. C., & Dawkins, M. P. (1979). Determinants of genocide fear in a rural Texas community: A research note. *American Journal of Public Health, 69,* 605–607.

Finnegan, J. R., Jr., Meischke, H., Zapka, J. G., Leviton, L., Meshack, A., Benjamin-Garner, R., et al. (2000). Patient delay in seeking care for heart attack symptoms: Findings from focus groups conducted in five U.S. regions. *Preventive Medicine, 31,* 205–213.

Fleming, D. T., & Wasserheit, J. N. (1999). From epidemiological synergy to public health policy and practice: The contribution of other sexually transmitted diseases to sexual transmission of HIV infection. *Sexually Transmitted Infections, 75,* 3–17.

France, D. (1998, December 22). Challenging the conventional stance on AIDS. *The New York Times*, p. F6.

Gamble, V. N. (1994). Race, class, and the pill: A history. In S. E. Samuels & M. D. Smith (Eds.), *The pill: From prescription to over-the-counter* (pp. 21–39). Menlo Park, CA: The Kaiser Family Foundation.

Gamble, V. N. (1997). Under the shadow of Tuskegee: African Americans and health care. *American Journal of Public Health, 87*, 1773–1778.

Gifford, A. L., Bormann, J. E., Shively, M. J., Wright, B. C., Richman, D. D., & Bozzette, S. A. (2000). Predictors of self-reported adherence and plasma HIV concentrations in patients on multi-drug antiretroviral regimens. *Journal of Acquired Immune Deficiency Syndromes, 23*, 386–395.

Henshaw, S. K. (1998). Unintended pregnancy in the United States. *Family Planning Perspectives, 30*, 24–29, 46.

Herek, G. M., & Capitanio, J. P. (1994). Conspiracies, contagion, and compassion: Trust and public reactions to AIDS. *AIDS Education and Prevention, 6*, 365–375.

Herek, G. M., & Glunt, E. K. (1991). AIDS-related attitudes in the United States: A preliminary conceptualization. *Journal of Sex Research, 28*, 99–123.

Janz, N. K., & Becker, M. H. (1984). The health belief model: A decade later. *Health Education Quarterly, 11*, 1–47.

Jones, J. H. (1993). *Bad blood: The Tuskegee syphilis experiment*. New York: The Free Press.

Kelly, J. A., Murphy, D. A., Sikkema, K. J., McAuliffe, T. L., Roffman, R. A., Solomon, L. J., et al. (1997). Randomised, controlled, community-level HIV-prevention intervention for sexual-risk behavior among homosexual men in U.S. cities. *Lancet, 350*, 1500–1505.

Kelly, J. A., St. Lawrence, J. S., Stevenson, L. Y., Hauth, A. C., Kalichman, S. C., Diaz, Y. E., et al. (1992). Community AIDS/HIV risk reduction: The effects of endorsements by popular people in three cities. *American Journal of Public Health, 82*, 1483–1489.

Klonoff, E. A., & Landrine, H. (1999a). Do Blacks believe that HIV/AIDS is a government conspiracy against them? *Preventive Medicine, 28*, 451–457.

Klonoff, E. A., & Landrine, H. (1999b). Cross-validation of the Schedule of Racists Events. *Journal of Black Psychology, 25*, 231–254.

Krieger, N. (2000). Discrimination and health. In L. F. Berkman & I. Kawachi (Eds.), *Social Epidemiology* (pp. 36–75). Oxford, UK: Oxford University Press.

Krieger, N., & Sidney, S. (1996). Racial discrimination and blood pressure: The CARDIA study of young black and white adults. *American Journal of Public Health, 86*, 1370–1378.

Landrine, H., & Klonoff, E. A. (1996). The schedule of racist events: A measure of racial discrimination and a study of its negative physical and mental health consequences. *Journal of Black Psychology, 22*, 144–168.

Landrine, H., & Klonoff, E. A. (2001). Cultural diversity and health psychology. In A. Baum, T. A. Revenson, & J. E. Singer (Eds.), *Handbook of health psychology* (pp. 851–891). Mahwah, NJ: Erlbaum.

LaVeist, T. A., Nickerson, K. J., & Bowie, J. V. (2000). Attitudes about racism, medical mistrust, and satisfaction with care among African American and white cardiac patients. *Medical Care Research and Review, 57*(Supp. 1), 146–161.

Lillie-Blanton, M., Brodie, M., Rowland, D., Altman, D., & McIntosh, M. (2000). Race, ethnicity, and the health care system: Public perceptions and experiences. *Medical Care Research and Review, 57*(Supp. 1), 218–235.

Lincoln, C. E. (1975). The Black masses and population control. *Health Education, 6*(5), 8–10.

Mayberry, R. M., Mili, F., & Ofili, E. (2000). Racial and ethnic differences in access to medical care. *Medical Care Research and Review, 57*(Supp. 1), 108–145.

McGary, H. (1999). Distrust, social justice, and health care. *The Mount Sinai Journal of Medicine, 66*, 236–240.

Parsons, S., Simmons, W., Shinhoster, F., & Kilburn, J. (1999). A test of the grapevine: An empirical examination of conspiracy theories among African Americans. *Sociological Spectrum, 19*, 201–222.

Piccinino, L. J., & Mosher, W. D. (1998). Trends in contraceptive use in the United States: 1982–1995. *Family Planning Perspectives, 30*, 4–10, 46.

Richardson, L. (1997, April 21). Experiment leaves legacy of distrust of new AIDS drugs. *The New York Times*, p. A1.

Roberts, D. (1998). *Killing the Black body: Race, reproduction, and the meanings of liberty*. New York: Vintage Books.

Roberts, D. (2000). Black women and the pill. *Family Planning Perspectives, 32*, 92–93.

Rodrigues-Triaz, H. (1982). Sterilization abuse. In R. Hubbard, M. S. Henifin, & B. Fried (Eds.), *Biological woman—The convenient myth* (pp. 147–160). Cambridge, MA: Schenkman Publishing Company, Inc.

Rothenberg, R. B., Wasserheit, J. N., St. Louis, M. E., Douglas, J. M., & the Ad Hoc STD/HIV Transmission Group. (2000). The effect of treating sexually transmitted diseases on the transmission of HIV in dually infected persons: A clinic-based estimate. *Sexually Transmitted Diseases, 27*, 411–416.

Safer, M. A., Tharps, Q. J., Jackson, T. C., & Leventhal, H. (1979). Determinants of three stages of delay in seeking care at a medical clinic. *Medical Care, 17*, 11–29.

St. Jean, Y., & Feagin, J. R. (1998). *Double burden: Black women and everyday racism*. Armonk, NY: ME Sharpe, Inc.

Sikkema, K. J., Kelly, J. A., Winett, R. A., Solomon, L. J., Cargill, V. A., Roffman, R., et al. (2000). Outcomes of a randomized community-level HIV prevention intervention for women living in 18 low-income housing developments. *American Journal of Public Health, 90*, 57–63.

Singh, N., Squier, C., Sivek, C., Wagener, M., Nguyen, M. H., & Yu, M. L. (1996). Determinants of compliance with antiretroviral therapy in patients with human immunodeficiency virus: Prospective assessment with implications for enhancing compliance. *AIDS Care, 8*, 261–269.

Smith, C. (1999). African Americans and the medical establishment. *The Mount Sinai Journal of Medicine, 66*, 280–281.

Smedley, B. D., Stith, A. Y., & Nelson, A. R. (Eds.). (2002). *Unequal treatment—Confronting racial and ethnic disparities in health care*. Washington, DC: The National Academy Press.

Stone, V. E., Mauch, M. Y., Steger, K., Janas, S. F., & Craven, D. E. (1997). Race, gender, drug use, and participation in AIDS clinical trials. *Journal of General Internal Medicine, 12*, 150–157.

Taylor, J. Y. (1999). Colonizing images and diagnostic labels: Oppressive mechanisms for African American women's health. *Advances in Nursing Science, 21*(3), 32–45.

Thomas, S. B., & Quinn, S. C. (1991). The Tuskegee Syphilis Study, 1932 to 1972: Implications for HIV education and AIDS risk education programs in the black community. *American Journal of Public Health, 81*, 1498–1504.

Thomas, S. B., & Quinn, S. C. (1993). The burdens of race and history on Black Americans attitudes toward needle exchange policy to prevent HIV disease. *Journal of Public Health Policy, Autumn 14*, 320–347.

Treadwell, M. (1972). Is abortion Black genocide? *Family Planning Perspectives, 4*, 4–5.

Turner, C., & Darity, W. A. (1973). Fears of genocide among Black Americans as related to age, sex, and region. *American Journal of Public Health, 63*, 1029–1034.

Turner, P. A. (1993). *I heard it through the grapevine: Rumor in African-American culture*. Berkeley, CA: University of California Press.

U.S. Census Bureau. (2001). *The Black population: 2000*. Retrieved December 1, 2004, from http://www.census.gov/prod/2001pubs/c2kbr01-5.pdf

van Ryn, M., & Burke, J. (2000). The effect of patient race and socio-economic status on physicians' perceptions of patients. *Social Science & Medicine, 50*, 813–828.

Weisbord, R. G. (1973). Birth control and the Black American: A matter of genocide? *Demography, 10*, 571–590.

SHERYL THORBURN BIRD, Ph.D, M.P.H., is Associate Professor in the Department of Public Health at Oregon State University. Her research interests include the prevention of HIV, other STIs, and unintended pregnancy; perceived discrimination and reproductive health; and the acceptability of reproductive technologies. She is Principal Investigator on a study funded by the National Institute of Child

Health and Human Development (NICHD) that examines whether conspiracy beliefs regarding HIV/AIDS and birth control are associated with high-risk sexual behaviors among African Americans of reproductive age.

LAURA M. BOGART, Ph.D., is a Social/Behavioral Scientist at RAND. Her research focuses on social cognition and health, including ethnic differences in health care attitudes and behaviors, adherence to HIV treatment, and prediction and measurement of risky sexual behavior. She is Co-Investigator on a NICHD-funded study that examines whether conspiracy beliefs regarding HIV/AIDS and birth control are associated with high-risk sexual behaviors among African Americans of reproductive age.

Journal of Social Issues, Vol. 61, No. 1, 2005, pp. 127–137

The Legal Aspects of Parental Rights in Assisted Reproductive Technology

John K. Ciccarelli*
Pasadena, California

Janice C. Ciccarelli
Claremont, California

This paper provides an overview of the different legal approaches that are used in various jurisdictions to determine parental rights and obligations of the parties involved in third party assisted reproduction. Additionally, the paper explores the differing legal models that are used depending on the method of surrogacy being utilized. The data demonstrates that a given method of surrogacy may well result in different procedures and outcomes regarding parental rights in different jurisdictions. This suggests the need for a uniform method to resolve parental rights where assisted reproductive technology is involved.

There are a plethora of legal and ethical conundrums that are presented as a result of third party assisted reproduction (Darr, 1999). The medical advances which have allowed infertile couples the opportunity to have children have greatly outpaced society's, and consequently the law's, ability to address the relationships and attendant rights and responsibilities which arise between the parties (Handel, Ciccarelli, & Hanafin, 1993).

In fact, there are new and novel legal issues that arise with surprising regularity. As examples, cases have been filed seeking to determine if frozen sperm could be left to one's girlfriend through a testamentary instrument such as a will (*Hecht v. Superior Court*, 1993); whether a surrogate can sue an agency for negligence when the child she carried is subsequently killed by the intended father (*Huddleston v. Infertility Center of America, Inc.*, 1992); who has the right to determine if

---

*Correspondence concerning this article should be addressed to John K. Ciccarelli, 225 S. Lake Avenue, Suite 300, Pasadena, California 91101 [e-mail: jciccarelli@earthlink.net].

cryo-preserved embryos will be used to create a baby where a couple has divorced after the embryos have been frozen but before they have been implanted (*Davis v. Davis*, 1992); can an intended parent escape liability for child support where a child has been conceived through third party assisted reproduction and the couple eventually divorces (*In re Marriage of Buzzanca*, 1998); and, can same sex partners legalize their parental rights to children born as a result of this new technology in a situation where an adoption by at least one putative parent will be required (*In re Adoption of RBF*, 2002).

These are but a few of the myriad and complex legal issues in assisted re-production for which the law has no well-established method of resolution. It is certain that many additional questions will need to be addressed by the courts or legislatures before this area of law can be considered settled. While the forego-ing issues are outside the scope of this paper, they are raised to give the reader a sense of how dramatically this new technology impacts on traditional concepts of parentage, family, the right to procreate, and other individual rights.

Before a resolution can be found to the more esoteric problems that arise, it is first necessary to address the most basic question; specifically, who is the legal mother and father of a child born through use of these new technologies. Unfortunately, the answer to this question is not always clear and the answer varies depending upon the state or country where the child is born. Anyone venturing into these muddy legal waters must exercise extreme cautions since most jurisdictions have not considered the issue. Of the jurisdictions that have dealt with these sticky matters, there has been no clear consensus as to what framework to apply to reach the ultimate decision (e.g., *Florida Statutes Annotated*, 2004; *In re Marriage of Moschetta*, 1994; *Johnson v. Calvert*, 1993; *Matter of Baby M*, 1988).

Before turning to the jurisdictional discrepancies, it is beneficial to define the terms that are commonly used to identify the various parties. Contractual parenting (commonly know as surrogacy) occurs when a couple, the intended parents, contracts with a woman to carry a child for them and to relinquish that child to them after birth (Ciccarelli, 1997; Ragone, 1996). There are two major types of surrogacy arrangements: traditional surrogacy and gestational surrogacy. In traditional surrogacy, the surrogate is impregnated with the sperm of the male partner of the intended parents through artificial insemination (AI). Therefore this is commonly referred to as AI surrogacy. In this case, the impregnated woman is both the genetic and birth (i.e., gestational) mother and the intended father is also the genetic father (Ciccarelli, 1997; Ragone, 1996). Gestational carrier surrogacy is used when the female partner of the intended couple has viable eggs but is unable to successfully carry a pregnancy to term. The intended mother's eggs are fertilized with her male partner's sperm in the laboratory using in vitro fertilization (IVF) and the embryo is then implanted in the "surrogate" mother's uterus. In gestational surrogacy, the woman who carries the child has no genetic connection to the child and the intended parents are also the genetic parents (Ciccarelli, 1997; Ragone,

1996). Based on the foregoing, this paper will use the term traditional surrogate for the woman who conceives via AI using the sperm of the father who intends to rear the child and the term gestational surrogate for the woman who carries an embryo that has been conceived via IVF using the intended couple's egg and sperm. The couple that contracts with the surrogate mother is referred to as the intended, social, commissioning, or contracting parents, depending on where they are in the surrogate parenting process.

In jurisdictions where the matter of surrogacy has been addressed, the end result runs the gamut of possibilities from declaring the entire activity illegal and void as against public policy (e.g., *Arizona Revised Statutes Annotated*, 2003; *Matter of Baby M*, 1988), to vesting legal rights with the intended parents (*Johnson v. Calvert*, 1993), to applying an adoption model wherein the surrogate mother is recognized as the parent who must then relinquish her parental rights to the intended parents (*A. H. W. v. P. W.*, 2000). Adding an overlying level of confusion to this entire process is the manner in which these laws have been established. Some jurisdictions have enacted legislation (see pages 132 and 133 for specific code sections) while others have simply allowed the courts to consider the outcomes where the legislature has failed to act.

The initial inquiry is almost always directed at the type of surrogacy that is involved. Specifically, was the child conceived through AI or IVF? This distinction has several legal ramifications. If the child is born as a result of AI then the court is often able to render a decision that is in the best interests of the child (*Matter of Baby M*, 1988). Conversely, in the case of IVF it is likely (depending on the jurisdiction) that the court will examine the situation in a manner more closely aligned with contract principles. Under such an analysis the court does not inquire into the best interests of the child (*Johnson v. Calvert*, 1993).

Courts that have addressed the issue of conflicting claims of parental rights where the child was conceived through AI have universally applied adoption law to resolve the dispute (*In re Marriage of Moschetta*, 1994; *Matter of Baby M*, 1988). In such a situation, since the surrogate is both genetically connected to the child as well as actually carrying the pregnancy, the courts have had little trouble reaching the conclusion that she is the legal mother of the child. Therefore, the logical extension of this conclusion is that she is in an analogous position to a birth mother in a traditional adoption. This means that the surrogate must relinquish her parental rights in order for the intended parents, and specifically the intended mother, to finalize her parental rights (*In re Marriage of Moschetta*, 1994; *Matter of Baby M*, 1988).

Under the adoption law of almost any jurisdiction pre-birth agreements to relinquish parental rights are deemed to be nugatory (e.g., *California Family Code*, 2004; *Florida Statutes Annotated*, 2004). The rationale for this is that a birth mother cannot make an informed decision regarding termination of her parental rights until after the child is born. Moreover, even after a birth mother has consented to relinquish her rights, she will have a period of time within which she can revoke her consent to an adoption (e.g., *California Family Code*, 2004). In practical terms, this means that any contract the intended parents enter into with an AI surrogate before the birth of the child will have no effect on her parental rights. This also means that the surrogate will have some amount of time, even after she has agreed to relinquish her rights, to change her mind. Not only does this place the couple at risk of not being able to finalize their parental rights through this process, but it also has more far reaching consequences as discussed below.

But what happens in the case where the surrogate abides by her agreement to relinquish her parental rights? The intended father is, essentially, in the same position as any man who impregnates a woman. He is entitled to a judgment of paternity, certainly after the birth of the baby and, in many jurisdictions, prior to the birth of the child. This then leaves finalization of the intended mother's parental rights. This is accomplished by undertaking and completing a step-parent adoption. A step-parent adoption is utilized since the intended mother is legally married to the intended and biological father (*In re Marriage of Moschetta*, 1994).

This fact is not without its own wrinkle. In such a situation the intended father is also a "natural" parent of the child who, in many jurisdictions, must consent to his wife's adoption of "his" child before the adoption can occur (see *California Family Code*, 2004). This fact can place the intended parents in an unequal bargaining position should conflict arise between them prior to completion of the adoption.

The outcome in a case where an AI surrogate refuses to adhere to the agreement is likely to engender litigation where a number of legal outcomes are possible. In the first instance, the surrogate may attempt to "cut off" any claim of parentage by the only person who could assert such a claim; namely, the intended, biological father. In order for such an assertion to prevail, it would be necessary to convince a judge that the intended father was nothing more than a sperm donor. As such he would have no parental rights with respect to any child born from the donation. To date, such an argument has been rejected by the courts regardless of whether the argument was made by the surrogate (to cut-off parental rights) or the intended father (often to circumvent child support obligations).

The more common result is for the court to find that the intended father and the surrogate are the "parents" of the child and treat the matter as a custody and visitation issue. Such a determination requires the court to consider the best interests of the child, particularly with regard to the custody arrangement (*In re Marriage of Moschetta,* 1994; *Matter of Baby M*, 1988). It is quite possible for this situation to result in a shared custody and visitation plan between the surrogate and the intended father with the attendant payment of child support. Needless to say, the intended mother would be without a method to finalize her parental rights absent the consent and cooperation of the surrogate.

The analysis undertaken by the courts is best illustrated in the two most well known cases involving an AI surrogate who changed her mind. These cases are *Matter of Baby M* (1988) and *In re Marriage of Moschetta* (1994). The *Baby M* decision was the first case to wrestle with the issue of parental rights in the context of a traditional surrogacy agreement. This case came about as a result of an agreement that was entered into on February 6, 1985 between the intended parents, William and Elizabeth Stern, and the surrogate, Mary Beth Whitehead. Baby M was born on March 27, 1986.

Although Mrs. Whitehead testified to developing a bond with the baby during the pregnancy, she became most vociferous immediately after birth. While exhibiting signs of an emotional crisis, Mrs. Whitehead, nonetheless, turned Baby M over to the Sterns on March 30, 1986. The next day she went to the Sterns home and told them how much she was suffering and that she could not live without the baby. The Sterns, fearful that she would commit suicide, gave the child to Mrs. Whitehead. The Sterns did not see the baby again until four months later when the baby was forcibly taken by the police from a home in Florida where the Whiteheads were hiding with her.

Mr. Stern filed a complaint seeking possession and custody of the child in addition to enforcement of the terms of the surrogacy contract. The trial court found that the surrogacy contract was valid, ordered Mrs. Whitehead's parental rights terminated, granted sole custody to Mr. Stern and entered an order allowing for Mrs. Stern to adopt the child without delay. An immediate appeal was taken and the ultimate result was diametrically opposite to the holding in the trial court.

The appellate court equated the surrogacy contract with baby selling and found it to be void and unenforceable as against public policy. The court found that payment to a surrogate was illegal, perhaps criminal and potentially degrading to women (*Matter of Baby M*, 1988). The court found, also, that the surrogate was the mother of the child, voided the trial court's termination of the surrogate's parental rights and the adoption of the baby by the intended mother. After undertaking an analysis of the circumstances that were in the best interests of the child, the Court granted custody to the natural father, Mr. Stern, and remanded the case to the lower court for a determination of the nature and extent of Mrs. Whitehead's visitation rights. Mrs. Stern was without any method for establishing parental rights absent the termination of Mrs. Whitehead's rights.

The central issue of the enforceability of a traditional surrogacy agreement was again taken up, this time by a California court, in the case entitled *In re Marriage of Moschetta*. In *Moschetta*, Robert and Cynthia Moschetta entered into an AI surrogacy agreement with Elvira Jordan in June or July of 1989. Ms. Jordan became pregnant in November of 1989. Unbeknownst to Ms. Jordan, the Moschettas began experiencing marital difficulties in January of 1990 and in April of that year Robert told Cynthia he wanted a divorce. Ms. Jordan became aware of these domestic issues while she was in labor on May 27, 1990. As a result of the couple's domestic difficulties, Ms. Jordan began to reconsider her agreement, but ultimately relented and allowed the Moschettas to take the baby from the hospital after they told her they were reconciled.

The marriage continued to deteriorate and on November 30, 1990 Robert moved from the home and took the baby with him. Cynthia filed for legal separation on December 21, 1990 and on January 11, 1992 she filed a petition to establish her parental rights vis-à-vis the baby contending that she was the *de facto* mother of the baby. In February, Ms. Jordan petitioned the court to be allowed to assert her rights by becoming a party in the dissolution action and her petition was granted in March.

The action was to be heard in three phases. The first was to determine the parental rights of Cynthia Moschetta and Elvira Jordan; the second was to determine custody and visitation; and the third was to conclude the marital dissolution. The Court found that Elvira Jordan was the "natural" mother of the baby because she not only gestated the child, but she was genetically connected to the baby. The Court also held that traditional surrogacy contract was unenforceable because, amongst other things, it contravened the state's adoption statutes by circumventing the formal consent to a child's adoption by the birth mother. Since the court adjudicated Mr. Moschetta and Ms. Jordan to be the parents of the baby, the only remaining question was the extent of legal and physical custody and visitation between these parties.

As these cases readily demonstrate, couples exploring the option of AI surrogacy cannot escape the legal risk associated with the procedure. They are in the same proverbial boat as those who elect to pursue a private adoption and are at risk for a period of time during which the birth mother can either refuse to relinquish her parental rights or change her mind after she has consented to the adoption.

If one is laboring under the assumption that the analysis is any more consistent in the case of IVF surrogacy a hefty dose of reality awaits them. The advantage to gestational surrogacy is that, in many cases, it allows the intended parents to obtain a pre-birth order declaring them the legal parents of the child (*Johnson v. Calvert*, 1993; *Belsito v. Clark*, 1994). While this offers a modicum of legal protection for those considering gestational surrogacy, once again, the outcome is based on the specific jurisdiction where the birth of the baby takes place (cf. *A. H. W. v. P. W.*, 2000).

As noted above, the majority of jurisdictions have not addressed the issue of surrogacy and the ones that have analyzed the issues have reached different results. California, Massachusetts and New Jersey have examined the issue through case law (*A. H. W. v. P. W.*, 2000; *In re Marriage of Buzzanca*, 1998; *In re Marriage of Moschetta*, 1994; *Matter of Baby M*, 1988; *Jaycee B. v. Superior Court*, 1996; *Johnson v. Calvert*, 1993; *R. R. v. M. H.*, 1998; *Smith v. Brown*, 1999;). Florida has enacted a statutory scheme to address surrogacy; the statute prohibits surrogates from being paid anything other than reasonable legal, living, and medical expenses (*Florida Statutes Annotated*, 2004). Washington, Louisiana, Nebraska and Kentucky statutorily prohibit surrogacy contracts that include any compensation to the surrogate (*Kentucky Revised Statutes Annotated*, 2002; *Louisiana Revised Statutes Annotated*, 2004; *Nebraska Revised Statutes*, 2003; *Washington Revised Codes Annotated*, 2004). New York, Utah, Michigan and Arizona ban surrogacy contracts as against public policy (*Arizona Revised Statutes Annotated*, 2003; *Michigan Comp. Laws Annotated*, 2003; *New York Domestic Relations Law*, 2004; *Utah Code Annotated*; 2004).

In Florida, the intended parents (commissioning couple) must file a petition with the court within three days after the birth of the child for an expedited affirmation of parental status. The surrogate, the doctor from the reproductive facility, and anyone claiming paternity are made aware of the hearing. At the hearing, the court determines the validity of the gestational surrogacy agreement and whether at least one of the commissioning parents is genetically connected to the child. Once these two matters are determined, the court issues an order directing the original birth certificate to be sealed and a new birth certificate to be prepared listing the commissioning parents as the legal parents (*Florida Statutes Annotated*, 2004).

Where the manner of establishing parental rights is left to the courts differing results have occurred. This is best demonstrated by examining the states of California and New Jersey. California allows the intended parents in an IVF surrogacy arrangement to apply for an order, prior to the birth of the baby, directing that they be recognized as the legal parents of the child carried by the surrogate and that their names to be placed on the birth certificate. In order to understand this result one must consider the California Supreme Court's holding in *Johnson v. Calvert* (1993).

In *Johnson*, the court dealt with the situation where an IVF surrogate, Anna Johnson, purportedly changed her mind and attempted to obtain custody of the child. The case started when Mark and Crispina Calvert contracted with Anna Johnson on January 15, 1990 to carry the embryo created with Mr. Calvert's sperm and Mrs. Calvert's egg. The relationship deteriorated between the parties for a number of reasons, but by July 1990, Ms. Johnson sent the Calverts a letter demanding payment or she would refuse to relinquish custody of the child. The Calverts responded by seeking court intervention.

A trial was held in October 1990 in which the parties stipulated to the fact that the Calverts were the genetic parents of the baby. The trial ended with a ruling that the Calverts were the "genetic, biological, and natural" father and mother and that Ms. Johnson had no parental rights or rights to visitation. Ms. Johnson appealed (*Johnson v. Calvert*, p. 88).

The California Supreme Court affirmed the trial court's decision in practical, but not legal terms. Specifically, the Court never addressed the issue of the enforceability of the surrogacy contract. Rather, the court analyzed the facts in light of the Uniform Parentage Act. Under this analysis they determined that Mr. Calvert was in, essentially, the same position as any father; namely, he had contributed his genetics to create a child. As to Ms. Johnson and Ms. Calvert, there was a tie with respect to maternity.

This tie came about because there is more than one way to establish maternity. One method is to contribute the genetic material that creates the child, and another is to carry the pregnancy. When these two methods do not coincide in the same woman, a tie is created. In order to break the tie, the Court examined the issue of intent. In other words, the salient inquiry became, who initiated the action that brought the child into existence. In making this determination the Court looked to the fact that there was a contract between the parties and, but for, the Calverts' intent there would have been no child.

The end result of this analysis is that the intended parents are the legal parents from the moment of pregnancy and as such are entitled to a pre birth order establishing this fact. California has extended the intent analysis even further and has prevented intended parents who were not genetically connected to the child (because the child was conceived from donated gametes) from renouncing parentage. (e.g., *Jaycee B. v. Superior Court*, 1996; *In re Marriage of Buzzanca*, 1998.)

The issue that has not been judicially resolved is whose rights prevail where a couple creates an embryo using donated sperm and egg, contracts with a surrogate to carry the embryo, and the surrogate subsequently changes her mind. Under the intent analysis of *Johnson* as extended in *Jaycee B* and *Buzzanca* an argument can be advanced that the intended parents are the legal parents because, but for their efforts, the child would not have come into being. Conversely, and specifically in the case of the intended mother, the argument could be made that there is no "tie" to break since the intended mother is not genetically connected to the child and the surrogate is gestating the baby. Accordingly, only the surrogate can establish maternity under this scenario. Needless to say, one may anticipate future litigation over this point.

The New Jersey Court reached a different result from California in *A. H. W. v. P. W.* (2000). In *A. H. W.* a couple contracted with the intended mother's sister to carry an IVF pregnancy for them. There was absolutely no dispute between the parties as to who were the legal parents of the child to be born. Rather, when the intended parents sought a declaration of maternity and paternity the Attorney General of the State of New Jersey opposed the order on the grounds that the requested relief was contrary to the law prohibiting surrender of a birth mother's rights until seventy-two hours after birth.

The Court first reviewed New Jersey's legal history in the law of surrogacy as set forth in *Matter of Baby M* and the public policy arguments regarding surrogacy. The court was mindful of the fact that the surrogate had no genetic ties to the child in this case, unlike Baby M. Nonetheless, in rejecting the intended parents' argument that the surrogate was analogous to an "incubator" the court responded as follows:

> While [the intended parents] are correct that [the surrogate] will have no biological ties to the baby, their simplistic comparison to an incubator disregards the fact that there are human emotions and biological changes involved in pregnancy.
>
> A bond is created between a gestational mother and the baby she carries in her womb for nine months. During the pregnancy, the fetus relies on the gestational mother for a myriad of contributions. A gestational mother's endocrine system determines the timing, amount and components of hormones that affect the fetus. The absence of any component at its appropriate time will irreversibly alter the life, mental capacity, appearance, susceptibility to disease and structure of the fetus forever. The gestational mother contributes an endocrine cascade that determines how the child will grow, when its cells will divide and differentiate in the womb, and how the child will appear and function for the rest of its life. (*A. H. W. v. P. W.*, 2000)

Not surprisingly, after the foregoing soliloquy, the court found that the gestational surrogate had seventy-two hours after the birth of the child before she could surrender her parental rights. During this time period she was legally vested with the right to make medical decisions on behalf of the child. The court did not address what other parental rights, if any, a gestational surrogate would have regarding a child born in such a situation. As the court observed, "That decision will have to be made if and when a gestational mother attempts to keep the infant after birth in violation of the prior agreement" *(A. H. W. v. P. W.*, 2000).

This leads to some very interesting and unanswered questions in New Jersey. For example, since the court recognized that the surrogate must wait seventy-two hours to surrender her rights, a priori, she must have rights to surrender. Conversely, the genetic parents, but for whose efforts the child would not have been created must have some rights. Is it then possible for the child to have three parents? In a situation where the surrogate changes her mind, whose rights does she cut-off, the intended mother or both intended parents? If the surrogate refuses to surrender her rights and has her name listed on the birth certificate, what happens in the event she later changes her mind. Is the intended mother forced to then complete an adoption? If so, how does one logically adopt one's own genetic offspring?

On a final note, there is another legal risk associated with surrogacy for the intended parents. Even if the surrogacy contract is executed in a jurisdiction that is favorable for surrogacy, there is nothing to prevent the surrogate from relocating to an unfavorable jurisdiction before the birth of the child. While there are no reported cases addressing such an event, it is quite probable that the law of the state where the birth of the child took place would govern the determination of parental rights (e.g., Weintraub, 1986).

There is no doubt that the emerging methods of reproduction are forcing us to re-examine traditional concepts regarding what constitutes family and parents. Absent of a national ban on surrogacy, it seems that this method of reproduction is here to stay. If this is so, and as the social and legal furor associated with cases of third party assisted reproduction clearly demonstrate, there is an immediate need for legislation in this area. This legislation should clearly define the parental rights and obligations of the parties involved in such an arrangement as well as the standards by which such a relationship could be initiated in the first place. The latter aspects should most certainly address the differences between AI and gestational surrogacy in addition to instituting safeguards to prevent the negative aspects of surrogacy, such as any economic coercion to the surrogate. In fact, no matter what one's perspective is on the ethics of surrogacy, it seems evident that any legislation must include a mechanism to examine the economic aspects in order to prevent the disenfranchisement of those of a lower socioeconomic status.

Moreover, this legislation should be drafted in a fashion analogous to the Uniform Parentage Act with a concerted campaign to convince each state legislature to adopt the model statute. (e.g., *California Family Code*, 2004.) Alternatively, the results as to parental rights will continue to vary from jurisdiction to jurisdiction. This will leave all involved parties in a precarious position because even if the agreement is made in a jurisdiction that is favorable to surrogacy, there is nothing that prevents one of the parties to the arrangement from relocating to an unfriendly jurisdiction.

References

A. H. W. v. P. W., 772 A.2d 948 (N.J. 2000).
Arizona Revised Statutes Annotated, § 25–218 (West 2003).
Belsito v. Clark, 644 N.E.2d 760 (Ohio 1994).
California Family Code, §§ 7600, *et. seq.*, §§ 8800, *et. seq.*, § 8801.3 and § 8814.5 (West 2004).
Ciccarelli, J. C. (1997). *The surrogate mother: A post-birth follow-up study*. Unpublished Doctoral Dissertation, California School of Professional Psychology, Los Angeles.
Darr, J. (1999). Assisted reproductive technologies and the pregnancy process: Developing an equality model to protect reproductive liberties. *American Journal of Law & Medicine, 25*, 455–476.
Davis v. Davis, 842 S.W.2d 588 (Tenn. 1992).
Florida Statutes Annotated, §§ 742.15, *et. seq.* and § 63.212 (West 2004).
Handel, W., Ciccarelli, J., & Hanafin, H. (1993). Legal and legislative aspects of gestational surrogacy. In R. H. Asch & J. W. W. Studd (Ed.), *Annual Progress in Reproductive Medicine* (pp.181–203). New York: The Parthenon Publishing Group.
Hecht v. Superior Court, 16 Cal.App.4th 836 (1993).
Huddleston v. Infertility Center of America, Inc., 700 A.2d 453 (Pa. 1997).
In re Adoption of RBF, 803 A.2d 1195 (Pa. 2002).
In re Marriage of Buzzanca, 61 Cal.App.4th 1410 (1998).
In re Marriage of Moschetta, 25 Cal.App.4th 1218 (1994).
Johnson v. Calvert, 5 Cal.4th 84 (1993).
Jaycee B. v. Superior Court, 42 Cal.App. 4th 718 (1996).
Kentucky Revised Statute Annotated, §199.590 (West 2002).
Louisiana Revised Statute Annotated, § 9:2713 (West 2004).
Matter of Baby M, 537 A.2d 1227 (N.J. 1988).
Michigan Comp. Laws Annotated, § 722.855 (West 2003).
Nebraska Revised Statutes, § 25-21, 200 (Michie 2003).
New York Domestic Relations Law, § 123 (LexisNexis 2004).
Ragone, H. (1996). Chasing the blood ties: Surrogate mothers, adoptive mothers and fathers. *American Ethnologist, 23*, 352–365.
R. R. v. M. H., 689 N.E. 2d 790 (Mass. 1998).

Smith v. Brown, 718 N.E. 2d 844 (Mass. 1999).
Utah Code Annotated, § 76-7-204 (LexisNexis 2004).
Washington Revised Code Annotated, §§ 26.26.210, *et. seq.* (LexisNexis 2004).
Weintraub, R. (1986). *Commentary on the conflicts of law* (4th ed.). St. Paul, MN: Thomson-West.

JOHN K. CICCARELLI is an attorney at law in private practice in Pasadena, California. He has extensive background in the field of reproductive law and ethics. He is a member of the Los Angeles County Bar Bioethics Committee and the American Bar Association Committee on Reproductive Law and Genetics.

JANICE C. CICCARELLI is a Clinical Psychologist in Private Practice in Claremont, California. She is an Adjunct Professor in the Psychology Psy.D. Department at the University of LaVerne. In her private practice, Dr. Ciccarelli counsels surrogates and couples involved in third party assisted reproduction.

138

Journal of Social Issues, Vol. 61, No. 1, 2005, pp. 139–157

Emergency Contraception: The Politics of Post-Coital Contraception

Christy A. Sherman*

Oregon Research Institute

The literature and events related to the politicization of emergency contraceptive pills (ECPs) in the United States is reviewed. The basis of opposition to the regimen, rooted in the mode of action of ECPs, the confusion with mifepristone, and the challenges this presents for ECP advocates is also discussed. Legislative actions that impact the availability of ECPs are described, as well as efforts to increase access and availability through innovative programs, legislation, and changes in medical practice. Recommendations for future research, service delivery, and public policy are also presented.

Emergency contraception has become increasingly politicized since the regimen received approval from the U.S. Food and Drug Administration (FDA) in 1997 (U.S. FDA, 1997). In the intervening period, several state legislative bodies have introduced and/or passed legislation impacting the availability of emergency contraceptive pills (ECPs) to U.S. women, including California, Illinois, Indiana, Kansas, Kentucky, Maryland, Minnesota, New York, Ohio, Virginia, and Washington (Center for Reproductive Law & Policy, 2001; Gellene, 2001; Kaiser Family Foundation, 2001a, 2001c, 2001d, 2001f, 2001h, 2002d; Long & Pearson, 2001; Smith, 2001; Wells et al., 1998). Legislative action with the potential to restrict access to ECPs has included refusal laws, or so-called conscience clause bills, and parental notification requirements for minors seeking ECPs. Legislative action with the potential to increase access to ECPs has included measures for pharmacy provision of ECPs and measures requiring provision of ECPs for victims of sexual assault. In addition to legislative activity, advocacy groups as well as professional associations have been lobbying both for and against increased availability of ECPs.

---

*Correspondence concerning this article should be addressed to Christy A. Sherman, Ph.D., Oregon Research Institute, Eugene, OR 97403 [e-mail: christys@ori.org].

ECPs consist of high doses of the hormones (estrogen and/or progestin) con-
tained in commonly used daily oral contraceptive pills (Hatcher et al., 1998).
Studies have shown ECPs can reduce a woman's risk of pregnancy from a single
episode of unprotected intercourse by at least 74% (e.g., Trussell, Rodriguez, &
Ellertson, 1999; von Hertzen et al., 2002; Yuzpe, Thurlow, Ramzy, & Leyshon,
1974). There are two types of ECPs on the market in the United States today,
one containing a combination of estrogen and progestin (Preven), known as the
Yuzpe regimen, and one containing only progestin (PlanB). The safety and effectiveness of the method in preventing pregnancy has been documented for over
25 years (e.g., Cheng, Gulmezoglu, Ezcurra, & Van Look, 2000; Norris Turner &
Ellertson, 2002; Trussell, Rodriguez, et al, 1999; Trussell & Stewart, 1992; von
Hertzen et al., 2002; Yuzpe & Lancee, 1977; Yuzpe, Smith, & Rademaker, 1982;
Yuzpe, Thurlow, et al., 1974; *Note.* In addition, the intra-uterine device [IUD] can
also be used for emergency contraception if inserted within five days of unprotected intercourse. See Hatcher et al., 1998 and Zhou & Xiao, 2001 for additional
information on use of the IUD for post-coital contraception).

However, due to a lack of a dedicated product until 1998, women have largely
been unaware of the regimen. Before 1998, women were able to utilize the method
only if they knew the number and type of oral contraceptive pills to take, or if they
participated in a demonstration project (Breitbart, Castle, Walsh, & Casanova,
1998; Petitti et al., 1998). Currently, ECPs are available in a variety of ways:
by prescription—either in person, over the phone, or over the internet; directly
through clinics (such as Planned Parenthood); and, in a handful of states, directly
through pharmacies engaged in collaborative prescribing agreements with health
care providers. In addition, several regular daily oral contraceptive pills can be
used (e.g., Lo-Ovral, Alesse). Although research is lacking on the numbers of
women who have utilized daily oral contraceptives for emergency contraception,
data reveal one emergency contraception (EC) information and referral Web site
(http://ec.princeton.edu), which provides instructions on which brands of oral contraceptives can be used and the dosage for each, received over 150,000 visits in
1998 and increased yearly to over 400,000 in 2002 (J. Trussell, personal communication, October 2, 2003).

Research on mode of action suggests that ECPs act to prevent pregnancy
in one of three ways depending upon when in the menstrual cycle the regimen
is taken: (a) delaying ovulation, (b) preventing fertilization, and (c) preventing
implantation of a fertilized egg (Croxatto et al., 2001; Hapangama, Glasier, &
Baird, 2001; Trussell & Raymond, 1999). Recent research suggests, however, that
prevention of implantation is not the primary mode of action (Marions et al., 2002;
Trussell, Ellertson, & Dorflinger, 2003), indicating the regimen may work more
like a traditional contraceptive (prior to fertilization) than previously thought.

The medical science related to the safety and effectiveness of emergency
contraception is clear and straightforward. The oral contraceptives used in ECP

products are a heavily studied class of medications and are non-addictive and non-toxic. ECPs have been declared safe and effective by the American College of Obstetricians and Gynecologists (1996, 2001a, 2001b), the U.S. Food and Drug Administration (U.S. FDA, 1997), the World Health Organization (WHO; WHO, 1998), in countless clinical trials (e.g., Arowojolu, Okewole, & Adekunle, 2002; Bagshaw, Edwards, & Tucker, 1988; Ellertson et al., 2003; Espinos et al., 1999; Raymond et al., 2000; Rodrigues, Grou, & Joly, 2001; von Hertzen et al., 2002), and in independent reviews of the research literature (e.g., Dunn et al., 2003; Norris, Turner, & Ellertson, 2002). In addition, emergency contraception is available in more than fifty nations worldwide (International Consortium for Emergency Contraception, n.d.). There are no contraindications to the use of ECPs because of the short duration of treatment, even for women who have contraindications to long-term use of hormonal contraceptives (International Planned Parenthood Federation, 2000; Webb & Taberner, 1993). Although the regimen can be taken up to 120 hours after sex (von Hertzen et al., 2002), the regimen is more effective the sooner it is initiated with higher levels of effectiveness when taken within 72 hours (Ashok et al., 2002; WHO, 1998), underscoring that timeliness of access is a critical issue.

Many hold high hopes for ECPs to reduce the number of unintended pregnancies because of their safety and effectiveness (Boonstra, 2002a, 2002b; Dailard, 2001; Glasier, 1998; Trussell et al., 1998; Trussell, Koenig, & Ellertson, 1997). While ECPs as a first-line contraceptive carry a higher risk of pregnancy than ongoing daily use of hormonal contraceptives, ECPs are being viewed by advocates as an important "back up" method to be used in case of contraceptive method failure. In a nation where one half of all pregnancies are unintended, and half of all unintended pregnancies end in abortion (Henshaw, 1998), additional effective methods to reduce unintended pregnancies are seriously needed. Data already suggest ECPs have decreased the number of abortions in the United States (Jones, Darroch, & Henshaw, 2002).

However, the potential of ECPs to drastically reduce the rate of unintended pregnancy in the United States is unlikely to be decided in a debate over the public health science, but in a political debate over the definitions of pregnancy, contraception, and abortion. The reason for the politicization of ECPs lies in interpretations of what it means for a method to prevent implantation of a fertilized egg. Advocates and opponents continue to debate the issue of where to place ECPs on the continuum of fertility control.

It is the point at which a fertilized egg has been implanted in the uterus that a woman is considered, medically, to be pregnant (American College of Obstetricians and Gynecologists, n.d.; Hughes, 1972; U.S. FDA, 1997). However, many opponents of ECPs disagree, and consider the moment of fertilization to be the moment when a human being is formed (Bossom, 2002; Wagner, 2001). Traditional oral contraceptive pills act primarily to prevent ovulation. This prevents fertilization as no egg is released (Hatcher et al., 1998). Medical abortifacients act following implantation (Spitz & Bardin, 1993). Because of their action both before and after fertilization, ECPs sit in a unique middle ground between traditional pre-coital contraception and early abortion. Opponents consider ECPs interference with the implantation of a fertilized egg to be abortion (Bossom, 2002; Wagner, 2001). Furthermore, because ECPs are a pharmacologic regimen potentially interfering with a fertilized egg, there has been confusion that ECPs are the same as mifepristone (Mifeprex), known as "the French abortion pill" (Harper & Ellertson, 1995; Jackson, Bimla, Schwarz, Freedman, & Darney, 2000).

It is important to clarify the differences between emergency contraceptive pills and medical abortion. Mifepristone received FDA approval in 2000 for use as an abortifacient (U.S. FDA, 2000a). Mifepristone is effective both prior to and following implantation (Kahn et al., 2000; von Hertzen et al., 2002), thus it can be used immediately following unprotected sex as well as up to several weeks following implantation (Schaff, Fielding, Eisinger, Stadalius, & Fuller, 2000). When taken following implantation, mifepristone will cause the implanted egg to stop developing, dislodge from the uterine wall and to be expelled from the uterus (Spitz & Bardin, 1993; U. S. FDA, 2000b). In contrast, ECPs are ineffective once implantation has occurred. ECPs will not interrupt, dislodge, or terminate an established pregnancy (WHO, 1998). For more information regarding mifepristone please see Baird (2000), Creinin (2000) and Harvey, Sherman, Bird, & Warren (2002).

Pro-choice advocates view ECPs as another contraceptive method because ECPs work prior to implantation (American Medical Women's Association, 1996; Planned Parenthood Federation of America, 2002). However, anti-choice advocates view it as an abortifacient because the regimen may, in some cases, work after fertilization (Bossom, 2002; Christian Medical Association, 1997; Wagner, 2001). It is not surprising that there has been confusion and that this regimen has been co-opted into the abortion debate. It is antiabortion advocates who have led the fight to restrict access and provision of ECPs, particularly to minors, by arguing that ECPs are abortifacients (e.g., American Life League, 1997; U.S. Conference of Catholic Bishops, 1998; Wagner, 2001). While the public health need for ECPs is clear, politics over abortion may prevent women from being able to access the method.

Efforts to Restrict ECP Availability

An example of administrative action to limit access to ECPs is found in a case in San Bernardino County, California. In March 2001 the San Bernardino County Board of Supervisors voted to ask for a federal waiver to ban the dispensing of emergency contraceptive pills in county clinics receiving federal funding for family planning services (Kaiser Family Foundation, 2001e; "A Reproductive Right," 2001). The "chief" concern cited by supervisors was the question of whether

ECPs prevent or terminate pregnancy (Gold, 2001a). The county's request was denied as it failed to meet the requirements of exceptional circumstances needed for a waiver (Gold, 2001b). The incident was, nonetheless, informative regarding objections to ECPs. Objections mirrored those often offered regarding availability of abortion services such as claims that the method is unsafe, concerns about side effects, beliefs that services should not be easily accessed by minors, that parents should be informed prior to providing services to minors, and that parents should be the party required to provide consent (Gold, 2001a). Objections to minors having access to ECPs reflect concerns that the method will result in increased sexual activity among teens (e.g., Golden et al., 2001).

Opponents nationwide have introduced bills at the state level that have the potential of limiting ECP availability. Nearly every state has some form of refusal law in place (Gold & Sonfield, 2000; Kaiser Family Foundation, 2001d). Refusal laws, sometimes referred to as conscience clause laws, purport to protect the rights of various health care workers to refuse to participate in providing services when those services conflict with their personal values and beliefs. Initially, the focus of these efforts was in allowing individuals to refuse involvement in services such as abortion. This effort has broadened to allowing not only providers and staff to refuse their participation, but institutions and employer-paid health plans as well (Gold & Sonfield, 2000; Querido, 1998). In the last few years refusal laws have attempted to secure the rights of providers, health care workers, and health care institutions to refuse to provide contraceptive-related services, particularly in health care facilities owned by religious organizations (Benjamin, 2002; Jones, 2003). Most recently this right of refusal has been extended to the provision of ECPs and even to the provision of information about emergency contraception (Gold & Sonfield, 2000).

While these laws may sound reasonable, their effect could be to reduce the ability of women to obtain reproductive health services (i.e., ECPs) in a timely fashion. Currently, this could happen in two ways. First, women who have experienced a contraceptive method failure, have not used contraception, or are victims of sexual assault might seek treatment for unprotected intercourse in a health care facility. However, if such a woman is treated in a health care facility that is exercising its right under a refusal law not to provide emergency contraception, she might never hear about ECPs. Smugar, Spina, & Merz (2000) found that in some hospitals women presenting to emergency departments following sexual assault learned of ECPs only if they asked, and that religious-owned hospitals exercising rights of refusal vary widely in their proscription of ECPs, ranging from policies prohibiting discussion of ECPs to policies requiring referrals for women to receive ECPs elsewhere. Thus, refusal laws can and do prevent women from being advised of a very effective option for avoiding pregnancy.

The second way refusal laws could impact access to ECPs lies in the gatekeeper role pharmacists can play with the potential to limit as well as increase access to

ECPs. Recently refusal laws have been extended to pharmacists under an expanded definition of those health care providers who hold right of refusal (Kaiser Family Foundation, 2002a). Thus, refusal laws can result in women being told that the pharmacist refuses to fill her prescription for ECPs. While in many areas a woman can simply go to another pharmacy, in some instances this will mean added delays to initiating treatment. Given the limited window within which the regimen must be initiated (most effective if used within 72 hours), this can pose significant problems and increase the risk of unintended pregnancy. In some areas (e.g., rural communities) there may not be another pharmacy to go to, meaning ECPs will be effectively unavailable.

There has also been legislative activity at the federal level to restrict access to ECPs. As in San Bernardino County and elsewhere, concerns about minors receiving ECPs have been echoed in legislation introduced in the U.S. Senate. Bills have been introduced requiring parental consent for ECPs in school-based health centers receiving federal funding, as well as motions to ban the distribution of ECPs in school clinics altogether (Kaiser Family Foundation, 2001i). While these measures have not yet been passed, it is likely that attempts to limit availability of ECPs to minors will continue.

Efforts to Increase ECP Availability

Supporters of access to ECPs have made some progress in making the regimen more available. Four major legislative and policy strategies have been used which have the potential of increasing women's access to ECPs: (a) laws that require health care facilities (e.g., emergency rooms) to inform women victims of sexual assault about ECPs (Kaiser Family Foundation, 2001c, 2001f, 2001h; Long & Pearson, 2001); (b) laws to allow provision of ECPs through pharmacists (Kaiser Family Foundation, 2001g, 2001j; Wells et al., 1998); (c) efforts to allow ECPs to be sold over-the-counter ("Chicago Women's Group," 2001; Kaiser Family Foundation, 2001a, 2001j), and (d) encouragement to health care providers to give advance prescriptions for ECPs to their patients (American College of Obstetricians and Gynecologists, 2001c).

As of March 2003 six states (California, Illinois, New York, Ohio, South Carolina, and Washington) had enacted provisions to require some form of emergency contraceptive service for survivors of sexual assault (Alan Guttmacher Institute, 2003). The Center for Reproductive Law and Policy reports that in 2002 eight additional states were considering legislation to require either health care facilities or law enforcement agencies to provide ECPs, or information about the regimen, to female victims of sexual assault (Arizona, Delaware, Florida, Hawaii, Illinois, Minnesota, New Jersey, and New York; Center for Reproductive Law & Policy, 2003), and others have considered bills in the past (e.g., Kaiser Family Foundation, 2001c, 2001f, 2001h). The controversy with this type of legislation is the impact such laws would have on hospitals owned and run by religious organizations, such as the Catholic church. While the larger issue of religious/private sector hospital mergers is beyond the scope of this article, the growing reality is that in some areas, religious-owned facilities are the sole provider of hospital-based care (such as emergency room services), and, in some areas, these hospitals have eliminated some types of reproductive health services such as emergency contraception following sexual assault (Donovan, 1996; Gallagher, 1997).

A law has been passed in New York state, without exclusions for facilities owned by religious organizations, requiring all hospitals to supply rape victims with ECPs (Sexual Assault Reform Act, 2003). A similar bill has been introduced in the state of Maryland (Kaiser Family Foundation, 2001c). While objections to the New York and Maryland bills were voiced by Catholic organizations (Kaiser Family Foundation, 2002b), a similar bill passed in Illinois with the support of the Illinois Catholic Conference (Reimer, 2002), suggesting variability in Catholic opposition. A recent survey found that 43% of Catholic-owned hospitals had policies prohibiting the discussion of emergency contraception with rape victims (Smugar, Spina, & Merz, 2000). However, some emergency room physicians at these facilities have gotten around the policy by informing women they cannot provide information about ECPs and giving a referral to a local provider who can prescribe the pills, or by writing a prescription on their personal prescription pad which does not include the hospital's name (Smugar, Spina, & Merz, 2000).

Some states, such as Alaska, California, and Washington, have passed laws allowing the dispensing of ECPs directly by pharmacists under cooperative prescribing agreements with local health care providers (Alan Guttmacher Institute, 2003; Gellene, 2001; Hutchings et al., 1998; Wells et al., 1998), eliminating the need to obtain a prescription. Efforts to pass similar laws have been under way in several other states (Hawaii, Minnesota, New Hampshire, New York, and Virginia; Center for Reproductive Law and Policy, 2002). Pharmacy provision has been heralded as a new avenue to increase access to ECPs for women (Boggess, 2002; Wells et al., 1998), simplifying what has been a cumbersome process of obtaining a prescription and filling it within a short period of time while the method can still be effective. Pharmacy provision has, in fact, increased the availability of ECPs to women (Sommers, Chaiyakunapruk, Gardner, & Winkler, 2001; Wells et al., 1998).

In February 2001 the Center for Reproductive Law and Policy filed a petition with the FDA to make ECPs available over the counter, without a prescription (Kaiser Family Foundation, 2001b). In April 2003 the makers of Plan B, the progestin-only ECP available in the United States, filed an application with the FDA to give Plan B over-the-counter status (Kaiser Family Foundation, 2003; Zernike, 2003). Five key professional medical organizations (American Medical Association, American Public Health Association, American Academy of Pediatrics, American Medical Women's Association, and American Society for

Reproductive Medicine) have all urged the FDA to grant ECPs over-the-counter status (American College of Obstetrics & Gynecology, 2001a; Foubister, 2001; "Over-counter," 2000). The American College of Obstetricians & Gynecologists issued a statement in February 2001 stating "We believe that emergency oral contraception can meet the FDA criteria for over-the-counter availability" (American College of Obstetrics & Gynecologists, 2001a, p. 1). Two years following the filing of the petition by Center for Reproductive Law and Policy the petition was still pending (Center for Reproductive Law and Policy, 2002; Center for Reproductive Rights, 2003).

In order for a drug to be granted over-the-counter status it must meet several criteria, including: low toxicity, no potential for overdose or addiction, no teratogenicity, no need for medical screening, self-identification of the need, uniform dosage, and no important drug interactions (Grimes, Raymond, & Scott Jones, 2001). ECP advocates argue the regimen meets all criteria (Ellertson, Trussell, Stewart, & Winikoff, 1998; Grimes et al., 2001). Ellertson et al. (1998) point out that clinicians rely upon women to determine their own need for ECPs, and that the only role for clinicians in diagnosing the need for ECPs would be to discourage a woman who has a low likelihood of pregnancy, although there is no "medically important reason" (Ellertson et al., 1998, p. 226) to discourage women on this basis. They note there is only one dosing schedule, regardless of whether a woman uses the Yuzpe regimen (Preven) or the levonorgestrel-only regimen (Plan B). They point out the strong safety record of the hormones used in ECPs and the fact that the drugs are one of the best scrutinized classes of medications in medicine. There is no risk of addiction, and, according to the WHO, no contraindications to use of the regimen with the exception of pregnancy, and only because the method is ineffective if the woman is already pregnant, not because of risks to the ongoing pregnancy (Raman-Wilms, Tsengt, Wighardt, Einarson, & Koren, 1995; WHO, 1998).

Those who support a change to over-the-counter status of ECPs cited the concern that this was the only way some women would have timely access to the method (e.g., Grimes, 2002; "Over-counter," 2000). Timely access to the method has implications for the effectiveness of the regimen. Data show that the sooner after unprotected intercourse the pills are taken the more effective ECPs are in preventing pregnancy (see Croxatto et al., 2001).

Providing women with a prescription in advance of an episode of unprotected intercourse would allow women to have the pills on hand should the need for the regimen arise, thereby circumventing obstacles that can introduce significant delays in initiating treatment. At the opening session of the 2001 annual meeting of American College of Obstetricians & Gynecologists the organization's new President called for members to provide advance prescriptions during routine gynecologic visits in an effort to reduce the rate of unintended pregnancy in the United States (American College of Obstetricians & Gynecologists, 2001c).

Provider and Consumer Acceptability

While earlier research found mixed reviews by health care providers about ECPs (Sherman, Harvey, Beckman, & Petitti, 2001a), it is of note that these studies were conducted prior to FDA approval. Initial provider concerns in one study in 1996 related primarily to lack of FDA approval and liability issues (Sherman et al., 2001a). In addition, a small percentage of providers believed the regimen to be an abortifacient (Sherman et al., 2001a). Providers in that same study, however, also demonstrated a lack of detailed knowledge regarding the method, including mode of action and contraindications (Sherman et al., 2001a). Nonetheless, there are data to suggest providers strongly support the idea of making ECPs available to women, including minors. Surveys conducted in the United States of providers, both before (Beckman, Harvey, Sherman, & Petitti, 2001; Delbanco et al., 1998; Kaiser Family Foundation, 1997; Sherman et al., 2001a; Sills, Chamberlain, & Teach, 2000) and after (Kaiser Family Foundation, 2003) FDA approval of ECPs, indicate a large proportion of providers are aware of the method and have positive attitudes toward it. A survey conducted in 1997 by the Kaiser Family Foundation revealed 100% of obstetricians/gynecologists and 98% of general practice providers reported ECPs to be safe, and 100% of providers surveyed reported the method was effective (Kaiser Family Foundation, 1997). Studies of health care providers have shown that even prior to FDA approval of ECPs, the majority of providers had positive attitudes regarding the regimen (Beckman et al., 2001; Delbanco et al., 1998; Kaiser Family Foundation, 1997; Sherman et al., 2001a; Sills et al., 2000).

In addition, the Kaiser Family Foundation conducted two surveys of obstetric/ gynecologic and family practice physicians in 1997 (Delbanco et al., 1998) and 2000 (Kaiser Family Foundation, 2000), before and after the FDA approval of ECPs. Compared to 1997, there were significant increases, in 2000, in the percent of physicians who discussed ECPs with their female patients "always" or "most of the time" and increases in the percent who had prescribed ECPs six or more times in the prior year (Kaiser Family Foundation, 2000). In addition, more physicians in 2000 felt there was increased interest in ECPs among their patients than had been the case in 1997 (Kaiser Family Foundation, 2000).

More recently, many professional medical organizations have called for increased availability and access, including advance prescription and over-the-counter status (American College of Obstetricians & Gynecologists, 2001b, 2001d; "Doc urges," 2001; "New ACOG Leader," 2001; "Over-counter," 2000). In February 2001, the Center for Reproductive Law and Policy (2001) filed a citizens petition with the FDA on behalf of over 70 medical and public health organizations calling for ECPs to be given over-the-counter status. In addition, in May 2001 a coalition of provider organizations, including the American College of Obstetricians and Gynecologists, the American Public Health Association, and the American

Academy of Pediatrics, lobbied to oppose efforts to require parental consent for provision of ECPs to minors (Kaiser Family Foundation, 2001i).

Data on consumer acceptability has shown very positive attitudes toward the method. In a survey of postpartum women, two-thirds reported they would be willing to use ECPs in the future (Jackson et al., 2000). In two surveys of actual users of the Yuzpe regimen, over 90% said they would use the regimen again (Breitbart et al., 1998; Harvey, Beckman, Sherman, & Petitti, 1999). Also, women reported high levels of satisfaction with the method (91%) and that they would recommend the method to family/friends (97%; Harvey et al., 1999).

In contrast to fears of opponents that the method would encourage having intercourse without use of contraception, most women surveyed in both the Breitbart et al. (1998) and Harvey et al. (1999) studies were using a method of contraception at the time the need for ECPs arose (90% and 70% respectively). Many women (84%) who had not been using a method of contraception previously were prompted by the need for ECPs to re-examine their birth control strategy and planned to begin using a regular method of contraception (Sherman, Harvey, Beckman, & Petitti, 2001b). What have been lacking are studies on the acceptability of ECPs among women from differing socio-cultural groups. Few studies with culturally diverse samples have specifically evaluated the relationship between culture (e.g., race/ethnicity) and acceptability among ECP users (e.g., Bird, Harvey, & Beckman, 1998). However, studies have documented high acceptability among women in a private health maintenance organization (Kaiser Permanente; Harvey et al., 1999) and in inner city clinics (Breitbart et al., 1998; Hakim-Elahi & Breitbart, 1998), suggesting women of varying income levels find the method acceptable.

While consumer acceptability of the method among users is high, the main obstacle to availability of the method is low awareness of the method that in turn results in low demand. Women's health advocates have been calling for increased public education about ECPs including awareness campaigns and increased patient counseling by health care providers (Boonstra, 2002a, 2002b). The American Public Health Association has adopted a public policy statement calling for promotion of ECPs among the public and health care providers, as well as eliminating barriers to access through protocols which ease dispensation of the regimen to women following unprotected intercourse (American Public Health Association, 2003).

Future Research

Research is needed regarding both providers and consumers of ECPs and public perceptions of ECPs. As advocates and opponents push for and against increased availability of ECPs, data are needed regarding the attitudes of providers and potential consumers to advance prescription, over-the-counter status, and pharmacy provision of ECPs. In addition, it is crucial to determine the impact of programs such as pharmacy projects on the ability of women to access this method within a limited time period. As additional modes of dispensing ECPs emerge, what will be the resulting impact upon access? Also, data regarding women's attitudes about the method and their preferred mode of service delivery are needed. Do women prefer an advance prescription, telephone prescription, or a face-to-face discussion with a health care provider? How is the method viewed by the general public? What perceptions—and misconceptions—do the public hold regarding the regimen? To what degree are ECP advocates successful in making the distinction between ECPs and mifepristone? The answers to these questions will inform women's health advocates and providers seeking to assist women in obtaining ECPs. The political meaning of ECPs deserves study also. Do women view the method as an abortifacient or as a contraceptive, and does this have an impact upon the acceptability of this method to potential consumers? Lastly, studies are needed that address cultural diversity; further study is needed regarding cultural differences in perceptions of the method among U.S. women including race/ethnicity, primary language, and socioeconomic status.

Recommendations for Service Delivery

Since endorsement of ECPs by the American College of Obstetricians and Gynecologists and its drafting of practice guidelines in 1996 (American College of Obstetricians and Gynecologists, 1996), ECPs have been considered the standard of care for treatment of women wishing to avoid pregnancy following unprotected intercourse. A sexual assault victim treated in a health care facility should, reasonably, be able to expect to be advised of all available treatment options, including ECPs, to prevent harmful sequelae of the assault. Failure to offer ECPs, or information about the regimen, to any woman who has experienced unprotected intercourse and wishes to avoid pregnancy could be characterized as practicing beneath the standard of care.

While it is important to meet the needs of women who are victims of sexual assault, women's health advocates point out that the reasons for needing emergency contraception are immaterial (Ellertson, Trussell, Stewart, Koenig et al., 2001). A powerful argument can be made that the woman who has had a contraceptive failure, or who has had unprotected intercourse for some other reason, is in no less need of the standard of care than the woman who has survived a sexual assault. Some have argued that health care providers who do not inform a woman who has been sexually assaulted, has had her contraceptive method fail, or has not used a contraceptive and wishes to avoid pregnancy have failed in their professional obligations to fully inform their patient of all available treatment options (Bell & Mahowald, 2000) and in their ethical responsibilities (Faundes, Brache, & Alvarez, 2003).

Health care providers of women of reproductive age need to become knowledgeable about ECPs and proactive in educating their female patients regarding the method. The regimen's safety and lack of contraindications makes this a method that can be prescribed widely and in advance. Providers should be prepared to offer all women of reproductive age information about the method and, for those women who are interested, an advance prescription. In areas where pharmacy opposition is known, providers should investigate the possibility of dispensing the pills directly from their offices to avoid a pharmacy visit.

Public Policy Recommendations

ECPs have become controversial because of the success of opponents to present the regimen as an abortion method. The debate over ECPs could mirror the contentious debate over abortion unless advocates are able to reframe the issue. Advocates need to present ECPs in a way that builds on the public health science of the regimen and capitalizes on its high level of provider and consumer acceptability. This must be accompanied by crafting policy and legislation that ensure timely access.

A challenge for ECP advocates in the United States will be to keep ECPs from being completely swallowed by the abortion debate. The ability of advocates to clearly articulate the mode of action of ECPs and to clarify its differences from medical abortion (e.g., mifepristone) may determine the ways in which the method is made available and the types of opposition that are mobilized. If ECPs are viewed by the public more as an abortifacient, then there are likely to be greater restrictions on availability and access. If ECP advocates are successful in promoting the regimen as a method which works prior to pregnancy, then it is likely that access to ECPs will increase through the spread of innovative pilot programs such as the pharmacy project in Washington state (Wells et al., 1998), and that lobbying for over-the-counter status will intensify.

ECP advocates will need to highlight that this regimen intervenes within hours of sexual intercourse if they are to be successful in gaining widespread public support. To advance their argument that this is a method which acts before pregnancy, ECP advocates will need to ensure that the public understands that, unlike methods of early abortion, ECPs act: prior to the point when a woman can test positive for pregnancy; for some women, before the egg has been ejected from the ovary into the fallopian tube; for some women, before the sperm has reached the egg; and, for some women, before a fertilized egg has implanted itself in the uterine wall and begun rapid cell division.

While emergency contraception needs to be distinguished from abortion, it is important to highlight that abortion is also a public health issue. Abortion helps reduce the number of unintended pregnancies, is associated with prevention of negative health outcomes for women, and is one of the most common and safe medical procedures women undergo (Finer & Henshaw, 2003; Kahn et al., 2000; Look, Mitchell, Rogers, Fox, & Lackie, 2002). Access to safe and legal abortion is as important a public health priority as access to emergency contraception.

We are in a historical time of rapid scientific discovery when intervention is possible earlier than ever before imagined. However, the question of ECP availability, like many other medical issues in the United States, has become highly politicized, and has somehow moved away from being a debate about the public health science of ECPs and the health and well being of women. Women's health advocates need to keep the focus on public health science. The problem of unintended pregnancy is a challenging one, and a promising regimen such as ECPs should be evaluated, not by the political meaning of its mode of action, but by its ability to safely reduce the number of unintended pregnancies and abortions among U.S. women.

References

Alan Guttmacher Institute. (2003, 1 March). *State policies in brief*. Retrieved March 28, 2003, from http://www.agi-usa.org/pubs/spib.html

American College of Obstetricians and Gynecologists. (n.d.). *Statement on contraceptive methods*. Retrieved March 30, 2003, from http://www.acog.org/from_home/departments/dept_notice.cfm?recno=11&bulletin=600

American College of Obstetricians and Gynecologists. (1996, December). Emergency oral contraception. *Practice Patterns*, Number 3.

American College of Obstetricians and Gynecologists. (2001a, February 14). *Statement of the American College of Obstetricians and Gynecologists supporting the availability of over-the-counter emergency contraception*. Retrieved April 8, 2002, from http://www.acog.org/from_home/publications/press_releases/nr02-14-01.cfm

American College of Obstetricians and Gynecologists. (2001b, February 28). *ACOG supports safety and availability of over-the-counter emergency contraception*. Retrieved April 8, 2002, from http://www.acog.org/from_home/publications/press_releases/nr02-28-01-2.cfm

American College of Obstetricians and Gynecologists. (2001c, April 30). *New ACOG leader promotes widespread advance prescriptions for emergency contraception*. Retrieved April 8, 2002, from http://www.acog.org/from_home/publications/press_releases/nr04-30-01-1.cfm

American College of Obstetricians and Gynecologists. (2001d, June 30). *FDA should grant OTC status to emergency oral contraception*. Retrieved April 8, 2002, from http://www.acog.org/from_home/publications/press_releases/nr06-30-01-3.cfm

American Life League. (1997). *Emergency contraception: The morning-after pill*. Retrieved March 30, 2003, from http://www.all.org/issues/bc05.htm

American Medical Women's Association. (1996). *Position statement on emergency contraception*. Retrieved September 21, 2003, from http://www.amwa-doc.org/index.cfm?objected-OEF88909_D567_OB25_531927/EE4CC23EFB

American Public Health Association. (2003). *Support of public education about emergency contraception and reduction or elimination of barriers to access*. Retrieved March 28, 2003, from http://www.apha.org/legislative/policy/policysearch/index.cfm?fuscaction=view&id=1252

Arowojolu, A. O., Okewole, I. A., & Adekunle, A. O. (2002). Comparative evaluation of the effectiveness and safety of two regimens of levonorgestrel for emergency contraception in Nigerians. *Contraception, 66*(4), 269–273.

Ashok, P. W., Stalder, C., Wagaarachchi, P. T., Flett, G. M., Melvin, L., & Templeton, A. (2002). A randomised study comparing a low dose of mifepristone and the Yuzpe regimen for emergency contraception for emergency contraception. *British Journal of Obstetrics and Gynecology,* *109*(5), 553–560.

Bagshaw, S. N., Edwards, D., & Tucker, A. K. (1988). Ethinyl oestradial and D-norgestrel is an effective emergency postcoital contraceptive: A report of its use in 1,200 patients in a family planning clinic. *The Australian and New Zealand Journal of Obstetrics and Gynaecology, 28*(2), 137–140.

Baird, D. T. (2000). Mode of action of medical methods of abortion. *Journal of the American Medical Women's Association, 55*(Suppl. 3), 121–126.

Beckman, L. J., Harvey, S. M., Sherman, C. A., & Petitti, D. B. (2001). Changes in providers' views and practices about emergency contraception: An HMO based intervention. *Obstetrics and Gynecology, 97*(6), 942–946.

Bell, B. S., & Mahowald, M. B. (2000). Emergency contraception and the ethics of discussing it prior to the emergency. *Women's Health Issues, 10*(6), 312–316.

Benjamin, E. (2002, April 9). Assembly passes women's health bill. *Albany Times Union*, p. B2.

Bird, S. T., Harvey, S. M., & Beckman, L. J. (1998). Emergency contraceptive pills: An exploratory study of knowledge and perceptions among Mexican women from both sides of the border. *Journal of American Medical Women's Association, 53*(5, Suppl. 2), 262–265.

Boggess, J. E. (2002). How can pharmacies improve access to emergency contraception? *Perspectives on Sexual and Reproductive Health, 34*(3), 162–165.

Boonstra, H. (2002a, October). Emergency contraception: The need to increase public awareness. *The Guttmacher Report, 5*(4), 3–6.

Boonstra, H. (2002b, December). Emergency contraception: Steps being taken to improve access. *The Guttmacher Report, 5*(5), 10–13.

Bossom, E. (2002, August 22). *Contraception or deception?* Retrieved September 21, 2003, from www.cwfa.org/articledisplay.asp?id=1559&department=CWA&categoryid=life

Breitbart, V., Castle, M. A., Walsh, K., & Casanova, C. (1998). The impact of patient experience on practice: The acceptability of emergency contraceptive pills on inner-city clinics. *Journal of the American Medical Women's Association, 53*(5, Supp 2), 255–257, 265.

Center for Reproductive Law and Policy. (2001). EC does it: Push for over-the-counter dispensing. *Reproductive Freedom News, 10*(2). Retrieved April 8, 2002, from http://www.crlp.org/rfn_01_03.html#ec

Center for Reproductive Law and Policy. (2002). Anniversary of CRLP petition to make EC available over-the-counter. *Reproductive Freedom News, 11*(2). Retrieved April 8, 2002, from http://www.crlp.org/rfn_02_02.html#EC

Center for Reproductive Law and Policy. (2003). *State trends in emergency contraception legislation.* Retrieved on April 8, 2002, from http://www.crlp.org/st_ec.html

Center for Reproductive Rights. (2003). *Two years later: Over the counter emergency contraception still stalled before Bush administration FDA.* Retrieved March 30, 2003, from http://www.reproductiverights.org/pr_03_0212ec.html

Cheng, L., Gulmezoglu, A. M., Ezcurra, E., & Van Look, P. F. (2000). Interventions for emergency contraception. *Cochrane Database of Systematic Reviews, 2*, CD001324.

Chicago women's group starts morning-after pill website. (2001, June 17). *Los Angeles Times*, A22.

Christian Medical & Dental Associations. (1997). *CMA Healthwise: "Morning-after pill."* Retrieved September 21, 2003, from http://www.cmdahome.org

Creinin, M. D. (2000). Medical abortion regimens: Historical context and overview. *American Journal of Obstetrics and Gynecology, 183*(Suppl. 2), S3–S9.

Croxatto, H. B., Devoto, L., Durand, M., Ezcurra, E., Larrea, F., Nagle, C., et al. (2001). Mechanism of action of hormonal preparations used for emergency contraception: A review of the literature. *Contraception, 63*(3), 111–121.

Dailard, C. (2001). Increased awareness needed to reach full potential of emergency contraception. *The Guttmacher Report, 4*(3), 4–5, 12.

Delbanco, S. F., Stewart, F. H., Koenig, J. D., Parker, M. L., Hoff, T., & McIntosh, M. (1998). Are we making progress with emergency contraception? Recent findings on American adults and health professionals. *Journal of the American Medical Women's Association, 53*(5, Suppl. 2), 242–246.

Doc urges emergency contraceptives. (2001, April 20). *The Los Angeles Times*. Retrieved April 8, 2002, from http://www.latimes.com/health/healthwires/20010430/tCBTOPAP.html

Donovan, P. (1996). Hospital mergers and reproductive health care. *Family Planning Perspectives, 28*(6), 281–284.

Dunn, S., Guilbert, E., Lefebvre, G., Allaire, C., Arneja, J., Birch, C., et al. (2003). Emergency contraception. *Journal of Obstetrics and Gyneacology Canada, 25*(8), 673–697.

Ellertson, C., Trussell, J., Stewart, F. H., Koenig, J., Raymond, E. G., & Shochet, T. (2001). Emergency contraception. *Seminars in Reproductive Medicine, 19*(4), 323–330.

Ellertson, C., Trussell, J., Stewart, F. H., & Winikoff, B. (1998). Should emergency contraceptive pills be available without prescription? *Journal of the American Medical Women's Association, 53*(5, Suppl. 2), 226–229, 232.

Ellertson, C., Webb, A., Blanchard, K., Bigrigg, A., Haskell, S., Shochet, T., et al. (2003). Modifying the Yuzpe regimen of emergency contraception: A multicenter randomized controlled trial. *Obstetrics and Gynecology, 101*(6), 1160–1167.

Espinos, J. J., Senosiain, R., Aura, M., Vanrell, C., Armengol, J., Cuberas, N., et al. (1999). Safety and effectiveness of postcoital contraception: A prospective study. *European Journal of Contraception and Reproductive Health Care, 4*(1), 27–33.

Faundes, A., Brache, V., & Alvarez, F. (2003). Emergency contraception: Clinical and ethical aspects. *International Journal of Gynaecology and Obstetrics, 82*(3), 297–305.

Finer, L. B., & Henshaw, S. K. (2003). Abortion incidence and services in the United States in 2000. *Perspectives on Sexual and Reproductive Health, 35*(1), 6–15.

Foubister, V. (2001, March 5). *OTC emergency contraceptives pushed, but not imminent: Medicine's widespread support for greater access to emergency contraception won't be enough to change its status from prescription only.* Retrieved April 8, 2002, from www.amednews.com/2001/prsc0305

Gallagher, J. (1997). Religious freedom, reproductive health care, and hospital mergers. *Journal of the American Medical Women's Association, 62*(2), 65–68.

Gellene, D. (2001, September 14). Lawmakers ok sales of morning-after pills. *Los Angeles Times*, p. B1.

Glasier, A. (1998). Safety of emergency contraception. *Journal of the American Medical Women's Association, 53*(5, Suppl. 2), 219–221.

Gold, S. (2001a, May 4). California and the west: Board rebuffed in its effort to ban morning-after pill. *Los Angeles Times*, p. A3.

Gold, S. (2001b, June 2). Agency rejects bid to ban pill. *Los Angeles Times*, p. B6.

Gold, R. B., & Sonfield, A. (2000). Refusing to participate in health care: A continuing debate. *The Guttmacher Report, 3*(6), 8–11.

Golden, N. H., Seigel, W. M., Fisher, M., Schneider, M., Quijano, E., Suss, A. et al. (2001). Emergency contraception: Pediatricians' knowledge, attitudes, and opinions. *Pediatrics, 107*(2), 287–292.

Grimes, D. A. (2002). Emergency contraception and fire extinguishers: A prevention paradox. *American Journal of Obstetrics and Gynecology, 187*(6), 1536–1538.

Grimes, D. A., Raymond, E. G., & Scott Jones, B. (2001). Emergency contraception over-the-counter: The medical and legal imperatives. *Obstetrics and Gynecology, 98*(1), 151–155.

Hakim-Elahi, E., & Breitbart, V. (1998). Experience and acceptability of emergency hormonal contraception. *Primary Care Update for Ob Gyns, 5*(4), 172.

Hapangama, D., Glasier, A. F., & Baird, D. T. (2001). The effects of peri-ovulatory administration of levonorgestrel on the menstrual cycle. *Contraception, 63*(3), 123–129.

Harper, C. C., & Ellertson, C. E. (1995). The emergency contraceptive pill: A survey of knowledge and attitudes among students at Princeton University. *American Journal of Obstetrics and Gynecology, 173*(5), 1438–1445.

Harvey, S. M., Beckman, L. J., Sherman, C. A., & Petitti, D. B. (1999). Women's experiences and satisfaction with emergency contraception. *Family Planning Perspectives, 31*(5), 237–240, 260.

Harvey, S. M., Sherman, C. A., Bird, S. T., & Warren, J. (2002). Understanding medical abortion: Policy, politics and women's health (Monograph). *Policy Matters, 3*, 1–50.

Hatcher, R. A., Trussell, J., Stewart, F., Cates, W., Stewart, G. K., Guest, F. et al. (1998). *Contraceptive Technology* (17th ed.). New York: Ardent Media Inc.

Henshaw, S. (1998). Unintended pregnancy in the United States. *Family Planning Perspectives, 30*(1), 24–29.

Hughes, E. C. (Ed.). (1972). Committee on terminology, The American College of Obstetricians and Gynecologists. *Obstetric-Gynecologic Terminology*. Philadelphia: Davis.

Hutchings, J., Winkler, J. L., Fuller, S., Gardner, J. S., Wells, E. S., Downing, D. et al. (1998). When the morning after is Sunday: Pharmacist prescribing of emergency contraceptive pills. *Journal of the American Medical Women's Association, 53*(5, Suppl. 2), 230–232.

International Consortium for Emergency Contraception. (n.d.). *ECPs Status and activity by country*. Retrieved November 29, 2004, from http://www.ceinfo.org/files/EC%20Status%20&%20Availability_09_29_03.pdf

International Planned Parenthood Federation. (2000). International Medical Advisory Panel statement on emergency contraception. *IPPF Medical Bulletin, 34*(3), 1–2.

Jackson, R., Bimla Schwarz, E., Freedman, L., & Darney, P. (2000). Knowledge and willingness to use emergency contraception among low-income post-partum women. *Contraception, 61*(6), 351–357.

Jones, M. (2003, June 5). Assembly bill on right of refusal passes. Milwaukee Journal Sentinel. Retrieved September 20, 2003, from http://www.jsonline.com/news/State/jun03/145785.asp

Jones, R. K., Darroch, J. E., & Henshaw, S. K. (2002). Contraceptive use among U.S. women having abortions in 2000–2001. *Perspectives on Sexual and Reproductive Health, 34*(6), 294–303.

Kahn, J. G., Becker, B. J., MacIsaa, L., Amory, J. K., Neuhaus, J., Olkin, I. et al. (2000). The effectiveness of medical abortion: A meta-analysis. *Contraception, 61*, 29–40.

Kaiser Family Foundation. (1997). *Kaiser Family Foundation national surveys of Americans and health care providers on emergency contraception*. Menlo Park, CA: Author.

Kaiser Family Foundation. (2000). Women's health care providers' experiences with emergency contraception. *Survey Snapshot*. Menlo Park, CA: Author.

Kaiser Family Foundation. (2001a, February 2). Virginia House votes in favor of allowing over-the-counter sale of emergency contraception. *Kaiser Daily Reproductive Health Report*. Retrieved March 20, 2002, from http://www.kaisernetwork.org/daily_reports/print_report.cfm?DR_ID=2620

Kaiser Family Foundation. (2001b, February 7). FDA to examine making Plan B emergency contraception available over the counter. *Kaiser Daily Reproductive Health Report*. Retrieved March 20, 2002, from http://www.kaisernetwork.org/daily_reports/print_report.cfm?DR_ID=2711

Kaiser Family Foundation. (2001c, February 13). Maryland House bill would require availability of emergency contraception for rape survivors. *Kaiser Daily Reproductive Health Report*. Retrieved March 20, 2002, from http://www.kaisernetwork.org/Daily_reports/rep_index.cfm?DR_ID=2819

Kaiser Family Foundation. (2001d, March 15). Several states considering "conscience" laws for pharmacists. *Kaiser Daily Reproductive Health Report*. Retrieved March 20, 2002, from http://www.kaisernetwork.org/Daily_reports/rep_index.cfm?DR_ID=3404

Kaiser Family Foundation. (2001e, March 20). San Bernardino County, California, requests ban on emergency contraception in public clinics, despite state and federal laws. *Kaiser Daily Reproductive Health Report*. Retrieved March 20, 2002, from http://www.kaisernetwork.org/Daily_reports/rep_index.cfm?DR_ID=3513

Kaiser Family Foundation. (2001f, April 10). New York bill would require hospital emergency rooms to offer emergency contraception to rape victims. *Kaiser Daily Reproductive Health Report*. Retrieved March 20, 2002, from http://www.kaisernetwork.org/Daily_reports/rep_index.cfm?DR_ID=3938

Kaiser Family Foundation. (2001g, April 11). California pilot program tests distributing emergency contraception without a prescription. *Kaiser Daily Reproductive Health Report*. Retrieved March 20, 2002, from http://www.kaisernetwork.org/Daily_reports/rep_index.cfm?DR_ID=3977

Kaiser Family Foundation. (2001h, May 7). Illinois House passes bill requiring hospitals to inform rape victims about EC, Governor "expected" to sign it. *Kaiser Daily Reproductive Health Report*. Retrieved March 20, 2002, from http://www.kaisernetwork.org/Daily_reports/rep_index.cfm?DR_ID=4446

Kaiser Family Foundation. (2001i, May 16). Health care organizations oppose legislation requiring parental consent for EC services at school-based health clinics. *Kaiser Daily Reproductive Health Report*. Retrieved March 20, 2002, from http://www.kaisernetwork.org/Daily_reports/rep_index.cfm?DR_ID=4629

Kaiser Family Foundation. (2001j, September 14). California lawmakers pass bill that would allow women to receive EC without a prescription. *Kaiser Daily Reproductive Health Report*. Retrieved March 20, 2002, from http://www.kaisernetwork.org/Daily_reports/rep_index.cfm?DR_ID=6941

Kaiser Family Foundation. (2002a, February 12). Virginia House passes late-term abortion ban, advances "feticide" and EC measures; Senate Committee rejects parental consent bill. *Kaiser Daily Reproductive Health Report*. Retrieved March 20, 2002, from http://www.kaisernetwork.org/daily_reports/print_report.cfm?DR_ID=9420

Kaiser Family Foundation (2002b, February 27). Maryland Catholic conference opposes bill requiring hospitals to inform rape, incest survivors about EC. *Kaiser Daily Reproductive Health Report*. Retrieved March 20, 2002, from http://www.kaisernetwork.org/daily_reports/print_report.cfm?DR_ID=9714

Kaiser Family Foundation. (2002c, April 2). Washington governor signs law requiring hospitals to offer emergency contraception to rape survivors. *Kaiser Daily Reproductive Health Report*. Retrieved April 8, 2002, from http://www.kaisernetwork.org/daily_reports/print_report.cfm?DR_ID = 10366

Kaiser Family Foundation. (2003, April 22). Emergency contraception maker applies to FDA for permission to sell pills over-the-counter. *Kaiser Daily Reproductive Health Report*. Retrieved September 29, 2003, from http://www.kaisernetwork.org/daily_reports/print_report.cfm?DR_ID=17279

Long, R., & Pearson, R. (2001, May 4). Lawmakers ok plan for rape victims. *Chicago Tribune*, p. A1.

Look, M. E., Mitchell, C. M., Rogers, A. J., Fox, M. C., & Lackie, E. G. (2002). Early surgical abortion: Efficacy and safety. *American Journal of Obstetrics and Gynecology, 187*(2), 407–411.

Marions, L., Hultenby, K., Lindell, I., Sun, X., Stabi, B., & Gemzell Danielsson, K. (2002). Emergency contraception with mifepristone and levonorgestrel: Mechanism of action. *Obstetrics and Gynecology, 100*(1), 65–71.

New ACOG leader promotes widespread advance prescriptions for emergency contraception. (2001, April 30). *American College of Obstetricians and Gynecologists*. Retrieved April 8, 2002, from http://www.acog.org/from_home/publications/press_releases/nr04-30-01-1.htm

Norris Turner, A., & Ellertson, C. (2002). How safe is emergency contraception? *Drug Safety, 25*(10), 695–706.

Over-counter 'morning-after' pill urged. (2000, December 6). *Los Angeles Times*, p. A22.

Petitti, D. B., Harvey, S. M., Preskill, D., Beckman, L. J., Postlethwaite, D., Switzky, H. et al. (1998). Emergency contraception: Preliminary report of a demonstration and evaluation project. *Journal of the American Medical Women's Association, 53*(5, Suppl. 2), 251–254.

Planned Parenthood Federation of America. (2002). *The difference between emergency contraception pills and medical abortion*. Retrieved September 21, 2003, from http://www.plannedparenthood.org/library/birthcontrol/ecandma.html

Querido, M. (1998). What are conscience clauses and how do they affect a woman's right to choose? *Reproductive Freedom News, 7*(7), 2–3.

Raman-Wilms, L., Tsengt, A. L., Wighardt, S., Einarson, T. R., & Koren, G. (1995). Fetal genital effects of first-trimester sex hormone exposure: A meta-analysis. *Obstetrics and Gynecology, 85*(1), 141–149.

Raymond, E. G., Creinin, M. D., Barnhard, K. T., Lovvorn, A. E., Rountree, W., & Trussell, J. (2000). Meclizine for prevention of nausea associated with emergency contraceptive pills: A randomized trial. *Obstetrics and Gynecology, 95*(2), 271–277.

Reimer, S. (2002, February 26). People should know about emergency contraception. *The Baltimore Sun*, p. 1E.

A reproductive right at risk: San Bernardino County's action against the morning-after pill will test just what limits the Bush White House will allow. (2001a, May 28). *Los Angeles Times*, p. B12.

Rodrigues, I., Grou, F., & Joly, J. (2001). Effectiveness of emergency contraception pills between 72 and 120 hours after unprotected sexual intercourse. *American Journal of Obstetrics and Gynecology, 184*(4), 531–537.

Schaff, E., Fielding, S. L., Eisinger, S. H., Stadalius, L. S., & Fuller, L. (2000). Low-dose mifepristone followed by vaginal misoprostol at 48 hours for abortion up to 63 days. *Contraception, 61*(1), 41–46.

Sexual Assault Reform Act, New York Stat. Ann. §2805-p, (2003).

Sherman, C. A., Harvey, S. M., Beckman, L. J., & Petitti, D. B. (2001a). Emergency contraception: Knowledge and attitudes of health care providers in a health maintenance organization. *Women's Health Issues, 11*(5), 448–457.

Sherman, C. A., Harvey, S. M., Beckman, L. J., & Petitti, D. B. (2001b, October). *Women's experiences and knowledge about emergency contraceptive pills.* Paper presented at the annual meeting of the American Public Health Association, Atlanta, GA.

Sills, M. R., Chamberlain, J. M., & Teach, S. J. (2000). The associations among pediatricians' knowledge, attitudes, and practices regarding emergency contraception. *Pediatrics, 105*(4 Part 2), 954–956.

Smith, T. (2001, February 21) "Morning after" pill approved senate version omits parental okay. *Richmond Times Dispatch,* p. A1.

Smugar, S. S., Spina, B. J., & Merz, J. F. (2000). Informed consent for emergency contraception: Variability in hospital care of rape victims. *American Journal of Public Health, 90*(9), 1372–1376.

Sommers, S. D., Chaiyakunapruk, N., Gardner, J. S., & Winkler, J. (2001). The emergency contraception collaborative prescribing experience in Washington State. *Journal of the American Pharmacological Association (Washington), 41*(1), 60–66.

Spitz, I. M., & Bardin, C. W. (1993). Clinical pharmacology of RU 486: An anti-progestin and antiglucocorticoid. *Contraception, 48*(5), 403–444.

Trussell, J. Bull, J., Koenig, J., Bass, M., Allina, A., & Gamble, V. N. (1998). Call 1-888-NOT-2-LATE: Promoting emergency contraception in the United States. *Journal of the American Medical Women's Association, 53*(5, Suppl. 2), 247–250.

Trussell, J., Ellertson, C., & Dorflinger, L. (2003). Effectiveness of the Yuzpe regimen of emergency contraception by cycle day of intercourse: Implications for mechanism of action. *Contraception, 67*(3), 167–171.

Trussell, J., Koenig, J., & Ellertson, C. (1997). Preventing unintended pregnancy: The cost-effectiveness of three methods of emergency contraception. *American Journal of Public Health, 87*(6), 932–937.

Trussell, J., & Raymond, E. G. (1999). Statistical evidence about the mechanism of action of the Yuzpe regimen of emergency contraception. *Obstetrics and Gynecology, 93*(5, pt. 2), 872–876.

Trussell, J., Rodriguez, G., & Ellertson, C. (1999). Updated estimates of the effectiveness of the Yuzpe regimen of emergency contraception. *Contraception, 59*(3), 147–151.

Trussell, J., & Stewart, F. (1992). The effectiveness of postcoital hormonal contraception. *Family Planning Perspectives, 24*(6), 262–264.

U.S. Conference of Catholic Bishops. (1998). Emergency Contraceptive Pills (ECPs): The truth, the whole truth, and nothing but the truth. *Life Insight, 9*(7). Retrieved March 31, 2003, from http://www.nccbuscc.org/prolife/publicat/lifeinsight/sept98.htm

U.S. Food and Drug Administration. (1997). Prescription drug products; certain oral contraceptives for use as postcoital emergency contraception. *Federal Register, 62,* 8610–8612.

U.S. Food and Drug Administration. (2000a, 28 September). *Approval letter Mifeprex (mifepristone) tablets.* Retrieved October 3, 2000, from http://www.fda.gov/cder/foi/appletter/2000/20687appltr.htm

U.S. Food and Drug Administration. (2000b). *Mifepristone questions and answers.* Retrieved January 15, 2001, from http://www.fda.gov/cder/drug/infopage/mifepristone/mifepristone-qa.htm

von Hertzen, H., Piaggio, G., Ding, J., Chen, J., Song, S., Bartfai, G. et al. (2002). Low dose mifepristone and two regimens of levonorgestrel for emergency contraception: A WHO multicentre randomised trial. *Lancet, 360*(9348), 1803–1810.

Wagner, T. R. (2001). Little pills: Targeting youth with new abortion drugs. *Insight*, 236. Retrieved March 30, 2003, from http://www.frc.org/get/is01i1.cfm

Webb, A., & Taberner, D. (1993). Clotting factors after emergency contraception. *Advances in Contraception, 9*(1), 75–82.

Wells, E. S., Hutchings, J., Gardner, J. S., Winkler, J. L., Fuller, T. S., Downing, D. et al. (1998). Using pharmacies in Washington state to expand access to emergency contraception. *Family Planning Perspectives, 30*(6), 288–290.

World Health Organization. (1998). *Emergency contraception: A guide for service delivery.* Geneva, Switzerland: Author.

Yuzpe, A. A., & Lancee, W. J. (1977). Ethinylestradiol and dl-norgestrel as a postcoital contraceptive. *Fertility and Sterility, 28*(9), 932–936.

Yuzpe, A. A., Smith, R. P., & Rademaker, A. W. (1982). A multicenter clinical investigation employing ethinyl estradiol combined with dl-norgestrel as postcoital contraceptive agent. *Fertility and Sterility, 37*(4), 508–513.

Yuzpe, A. A., Thurlow, H. J., Ramzy, I., & Leyshon, J. I. (1974). Post-coital contraception: A pilot study. *Journal of Reproductive Medicine, 13*(2), 53–58.

Zernike, K. (2003, April 22). Morning-after pill. *New York Times*, p. A17.

Zhou, L., & Xiao, B. (2001). Emergency contraception with multiload Cu-375 SL IUD: A multicenter clinical trial. *Contraception, 64*(2), 107–112.

CHRISTY SHERMAN is Associate Research Scientist at the Oregon Research Institute, and a Research Affiliate with the University of Oregon, Research Program in Women's Health, in Eugene, Oregon. She currently conducts research in women's reproductive health, specifically the reduction of unintended pregnancy and STIs among diverse populations of men and women. She is currently principal investigator on a project funded by the Centers for Disease Control and Prevention to develop and evaluate a program to increase partner notification among STD clients in public and private clinic settings. She was Project Director for the nation's first large-scale demonstration project in emergency contraception conducted in collaboration with Kaiser Permanente of Southern California in 1996–1998. She holds a doctorate in clinical psychology.

Journal of Social Issues, Vol. 61, No. 1, 2005, pp. 159–179

When Practices, Promises, Profits, and Policies Outpace Hard Evidence: The Post-Menopausal Hormone Debate

Michelle J. Naughton,* **Alison Snow Jones, and Sally A. Shumaker**
Wake Forest University School of Medicine

Currently, there is widespread controversy regarding the risks and benefits of hormone therapy for women over 50. The history of hormone therapy provides an excellent example of how different constituencies with competing objectives can produce health practices and policies of questionable benefit. We examine this history from the perspectives of women *who now live longer, expecting higher quality of life throughout their later years,* healthcare providers *who are influenced by the real and perceived needs of their patients as well as information provided by drug manufacturers, the* pharmaceutical industry *which seeks to identify and promote drugs that offer the most promise for both patients and shareholders, and* medical researchers—*including the National Institutes of Health and the Federal Drug Administration.*

In contrast to the other articles in this issue, we address the use of estrogens and progestins *after* women cease to produce these hormones on their own (endogenously)—that is, women's "older" postmenopausal years. The use of exogenous estrogens, and more recently progestins, for women in their peri- and postmenopausal years has gone through a series of transformations over the past 100 years. The history of these drugs serves as an excellent example of how different constituencies with competing objectives often collide and produce health practices and policies with questionable benefits and potential harm. Interestingly, in this complex historical drama, the "patients," women in their fifties and beyond, emerge as both willing participants and unwilling victims of the unintended consequences of these drugs.

_____
*Correspondence concerning this article should be addressed to Michelle J. Naughton, Ph.D., Department of Public Health Sciences, Wake Forest University School of Medicine, Medical Center Boulevard, Winston-Salem, NC 27157-1063 [e-mail: naughton@wfubmc.edu].

In the following pages we consider the history of menopause and hormone therapy (HT) from the perspectives of the various constituencies or key stakeholders including women who are living longer and expecting (possibly demanding) a higher quality of life throughout their later years, moving into and staying in the workplace at higher proportions and higher levels of responsibility, and more openly discussing topics like sexuality and menopause; the health providers who are influenced by the real and perceived needs of their patients as well as the information channeled to them from many sources, including the pharmaceutical industry; the pharmaceutical industry as it tries to identify and promote those drugs that offer the most promise for both the patient population and their shareholders; the medical research world—dominated most by the National Institutes of Health; the regulatory agencies (primarily the Food and Drug Administration); and the policymakers.

Hormone Therapy and Menopause

Menopause refers to the cessation of menstrual periods and the functions of the ovaries in women. The average age of menopause is approximately 51, with a range from 47–55 years (McKinlay, Brambilla, & Posner, 1992). The menopausal process is not a discrete event for most women, but rather a gradual process that occurs over a period of years as hormone levels involved in ovulatory cycles fluctuate and estrogen levels decline (Gannon, 1990). Most commonly, if a woman has not had a menstrual period for 12 months, she is considered to be postmenopausal (Schmidt & Rubinow, 1991), and 95% of women complete menopause by age 55.

Throughout history, a subset of women has survived childhood diseases and childbearing to live a relatively long and healthy life beyond the menopause. However, the number of women making it to their "years of wisdom" was relatively low. In the year 1900, the life expectancy of White and Black women in the United States was 49 and 34 years, respectively (National Center for Health Statistics, 1999). Currently, the life expectancy for White and Black women is approximately 77–80 years. Thus, today, an average woman in the United States can expect to live 35%–40% of her life postmenopausally. With the current aging of the U.S. population and the large post World War II "baby boom" generation, roughly 45 million women in the United States are now postmenopausal (*U.S. Census Projections*, 2003).

Common symptoms during the menopausal period include hot flushes, night sweats, vaginal dryness, sleep problems, and mood disturbances (Achilles & Leppert, 1997) The latter two symptoms, sleep problems and mood disturbances, may be due to hot flushes and night sweats, rather than being independent outcomes (Avis, Crawford, Stellato, & Longcope, 2001). The degree to which these symptoms bother women varies substantially in terms of both severity and the ability of the woman to tolerate such symptoms in her daily life (Achilles & Leppert, 1997). A relatively small portion of women experience no symptoms—other than the cessation of their menstrual cycle—during and after menopause. Others experience severe symptoms that seriously interfere with their ability to function on a day-to-day basis. The vast majority of women, however, are not at these extremes. Rather, many women experience some vasomotor symptoms (e.g., night sweats and hot flushes) for a period ranging from a few months to a couple of years around the menopause (Achilles & Leppert, 1997).

Menopause and its symptoms have been documented to occur in human females for centuries (Amundsen & Diers, 1973; Utian, 1980). In some time periods, menopause was seen as conferring on women renewed status and vigor, as the freedom from childbearing was achieved (Benedek, 1950). In other times, however, menopause was viewed as a mark of old age, disease, mental afflictions, and the lack of usefulness in society (Deutsch, 1945; Swartzman & Leiblum, 1987). A variety of "treatments" emerged to treat menopausal women over time, such as herbal remedies, aromatherapy, and reproductive surgeries, although hormonal therapy has been the predominant form of treatment for most of the twentieth century (Healy, 1995).

Hormones, synthetic and otherwise, have been available for roughly 75 years in the United States. Premarin is the most common oral hormone preparation used by postmenopausal women in the United States. It is comprised of conjugated estrogens derived from mare's urine during the third through tenth month of the equine gestational period. Premarin (PREgnant MAre urINe), was first developed and marketed by Ayerst, now Wyeth, in 1942 (Spake, 2002). Aggressive marketing of Premarin for the relief of menopausal symptoms was documented by 1945 (Seaman, 2003). Pharmaceutical company Schering AG synthesized ethinyl estradiol, subsequently used in birth control pills, during World War II. It was approved for human use in 1949 (Seaman, 2003).

As early as 1932, data suggested that estrogens induced mammary cancer in mice (Seaman, 2003). In 1939, the *Journal of the American Medical Association* published editorials urging a thorough investigation of diethylstilbestrol (DES), a synthetic hormone, prior to approval by the Food and Drug Administration (FDA) because of the possibility of carcinogenesis (MacBryde, Freedman, & Loeffel, 1939; Buxton & Engle, 1939; Shorr, Robinson, & Papanicolaou, 1939). Despite this knowledge, the FDA approved, in 1941, the use of stilbestrol for the treatment of menopausal symptoms and menstrual disorders. Evidence suggests approval resulted largely from concerted lobbying by drug companies aided by the cooperation from physicians enlisted in industry's cause and, presumably, interested in providing relief to those women experiencing severe menopausal symptoms (Seaman, 2003, p. 45). Estrogens proved to be highly effective in treating peri- and postmenopausal symptoms. However, little was known regarding appropriate dosages, and potential short- and long-term risks and benefits.

During the 1950s and 1960s, estrogen therapy gained popularity in the treatment of menopausal symptoms. In addition, and without research evidence, estrogens began to be marketed as a way in which women could retain a more youthful appearance and stave off the effects of the aging process. Wilson (1966) espoused this sentiment most forcefully in his book, *Feminine Forever*, which was widely disseminated and quoted during the 1960s and early 1970s. Wilson was among the first to portray menopause as a deficiency disease, comparable to diabetes, and requiring treatment. He asserted that estrogen, aside from keeping a woman sexually active and interested, maintained the strength of a woman's bones, the glow and firmness of her skin, and returned her to her previous balanced psychic functions, which made her more even tempered and agreeable.

In 1969, Reuben (1969) published "*Everything You Always Wanted to Know About Sex*," in which he wrote: "As the estrogen is shut off, a woman comes as close as she can to being a man . . . . To many women, the menopause marks the end of their useful life" (pp. 292–293). Reuben's book was followed by articles in women's magazines touting the wonders of hormone therapy. Excerpts from a 1973 *Harper's Bazaar* article, for example, suggest that the fountain of youth is at hand: "There doesn't seem to be a sexy thing estrogen can't and won't do to keep you flirtatiously feminine . . . . Prevalent medical opinion is that the safety and benefits of ERT (estrogen replacement therapy) have been convincingly demonstrated" (as cited in Spake, 2002, p. 3). These claims were made despite the fact that endometrial and breast cancer had been documented by that time as adverse effects of using estrogen therapy (Grady et al., 1992).

Based on short-term randomized clinical trials, there is clear evidence that estrogen therapy can mitigate the effects of menopausal symptoms, such as night sweats, hot flushes, sleep disturbance and vaginal dryness (Barnabei et al., 2002; Greendale et al., 1998; MacLennan, Lester, & Moore, 2001). In fact, symptom relief is the primary indication for which estrogen therapy was approved by the FDA in 1942 and it remains the most frequent reason why estrogen therapy is initiated in early postmenopausal women. Estrogen therapy is also routinely given to the majority of pre-menopausal women who undergo a hysterectomy (removal of the uterus) and bilateral oophorectomy (removal of the ovaries). Such women are cast into an immediate and abrupt menopause and, depending on their age at the time of surgery, may experience severe menopausal symptoms.

As the use of estrogen therapy increased, physicians began to note a rise in endometrial cancer, a cancer of the uterine lining. By the late 1970s, this increased risk was clearly documented (Grady et al., 1992), and the use of estrogens declined for women who had a uterus. By the mid-1980s, however, lower dose estrogen plus progestin preparations were introduced into the market, in order to address the increased risks of endometrial cancer among women who had not had a hysterectomy and who wanted to use estrogen. With these changes, prescriptions for estrogen alone (for those women without a uterus), and estrogen with progestin (for those women who had not had a hysterectomy) began to rise again. At last, women apparently had available to them safe hormone therapies that effectively addressed the symptoms associated with menopause.

Hormone Therapy and Other Health Conditions

During the 1980s, industry and researchers began to investigate whether hormone therapy bestowed benefits that went beyond the treatment of menopausal symptoms. Data began to emerge suggesting a benefit for osteoporosis—a debilitating and costly disease that disproportionately affects older women. Studies supported the benefits of HT for the prevention of bone loss (see, e.g., Grady et al., 1992), and in the 1990s, FDA approval was given for the use of HT in the prevention and treatment of osteoporosis. Recent data from the largest hormone therapy clinical trial on women and bone disease, the Women's Health Initiative Hormone Trials, demonstrate the benefits of estrogen with and without progestin, on the reduction of fractures in postmenopausal women (The Women's Health Initiative Steering Committee, 2004; Writing Group for the Women's Health Initiative Investigators, 2002).

Hormone therapies have been strongly advocated, also, by pharmaceutical companies and physicians for the primary and secondary prevention of heart disease over the past 25 years and, more recently, for the prevention and treatment of cognitive decline and dementia. Thus, what began as a therapy for the bothersome symptoms that arise around menopause and dissipate within a couple of years for most women became a potential long-term therapy for the prevention and treatment of three of the most common causes of morbidity and mortality in women: heart disease, osteoporosis, and dementia. How did this happen?

It has long been noted that women are at lower risk for premature heart disease than men—with women's risk rising after the menopause (Robinson, Cohen, & Higano, 1958; Stampfer & Colditz, 1991). Also, women appear to be at higher risk for dementia than men (Birge, 1997; Rapp et al., 2003; Shumaker et al., 2003). Interest began to emerge in the potential role that estrogen might play in this gender difference in heart disease and dementia risk profiles. As a consequence, observational studies (both cross sectional and longitudinal) began to appear in which the relationships between prior hormone use and heart disease and dementia were explored. For the most part, these studies, coupled with animal models, provided support for the hypothesis that HT might prevent and slow the progression of heart disease (Grady et al., 1992) and dementia (Nelson, Humphrey, Nygren, Teutsch, & Allan, 2002; Resnick, Maki, Golski, Kraut, & Zonderman, 1998; Yaffe, Sawaya, Lieberburg, & Grady, 1998). In order to provide protection from heart disease, dementia, and osteoporosis, many physicians began to believe that postmenopausal women should initiate hormone therapy at or near menopause and remain on hormone replacement therapy for the duration of their lives. Further, based on these presumed benefits of estrogen or HT, women well past menopause were being encouraged to initiate (or re-initiate) hormone therapy. As a consequence, during the late '80s and into the '90s, there was a marked rise in the promotion of hormone therapy among all menopausal women with a concomitant increase in prescription use (Kreling, Mott, Widerholt, Lundy, & Levitt, 2001).

Postmenopausal hormones emerged as one of the most commonly prescribed drugs in the United States—a prescription pattern that was initiated and sustained "off label." That is, HT was being widely advertised, prescribed, and endorsed for uses that were not approved by the FDA. This major trend in practice patterns concerned officials at the FDA. In 1999, for example, the FDA's Division of Drug Marketing, Advertising, and Communications (DDMAC) sent the following letter to what was then Wyeth-Ayerst (the company that owned Premarin—the most widely used estrogen in the United States):

> DDMAC is concerned that Wyeth-Ayerst continues to promote broad and ambiguous health claims for Premarin that promise yet-to-be substantiated or even identified health benefits from the use of Premarin. This is particularly troublesome in light of the prominent boxed warnings and numerous contraindications to the use of Premarin, and the serious risks, particularly long-term, associated with the use of Premarin. (Untitled Letter, January 11, 1999, p. 3)

The reason for this warning was the predominant lack of randomized, placebo controlled clinical trials regarding the use of hormone therapy for the primary and secondary prevention of heart disease or dementia. In examining the literature on heart disease, for example, most of the evidence regarding the protective effect of estrogen or HT against heart disease comes from observational or matched case-control studies. Such studies are likely to yield biased results when compared to results from randomized (and double masked) clinical trials or experimental designs (Friedman, Furberg, & DeMets, 1996). In the latter type of design, participants are randomly assigned to receive specific treatment(s) in order to reduce bias and a placebo control group is included for comparison purposes. In masked trials, participants do not know to which treatment group they have been assigned (single masked) and the study investigators also are masked to the participants' treatment group assignment (double masked). In observational studies, women have chosen to take hormone therapy. Bias can occur because women who choose to use hormone therapy rather than being randomized to receive it also tend to see a physician regularly, are in better health at drug initiation and are at lower risk for disease, have access to health care, and are generally better educated. In addition, women who choose to use estrogen may have generally healthier lifestyles and, consequently, less underlying disease than those who do not use HT (Grodstein & Stampfer, 1995). The majority of users are also Caucasian women, so the effects of HT on minority women have been understudied. Moreover, few clinical trials had examined the effects of HT on women across a range of age groups (ages 50–80+ years) for long term use. All of these factors complicate the evaluation of the role of HT in women's health, particularly when they are associated with a woman's decision to use HT.

Findings from all of the observational or non-randomized epidemiologic studies were not completely consistent and some clear risks from HT were emerging— for example, endometrial and breast cancer. In 1992, Grady and her colleagues (1992) performed a meta-analysis of studies published since 1970 that examined the effect of estrogen therapy and estrogen plus progestin therapy on endometrial cancer, breast cancer, coronary heart disease, osteoporosis, and stroke. They found evidence that estrogen therapy decreased the risk for coronary heart disease and for hip fracture, but long-term use increased the risk for endometrial cancer and was possibly associated with a small increase in breast cancer. The increase in endometrial cancer was thought to be diminished when progestin was added to the estrogen regimen in women whose uteri had not been removed. However, these researchers noted that the effects of combination hormones on the risk for other diseases had not been studied adequately at that time.

Recent Research Results on HT

Based on clear changes in practice patterns and emerging data suggesting that there may be risks associated with hormone therapy, some women's groups, the National Institutes of Health (NIH), the FDA and, eventually, the U.S. Congress began to demand research that would provide clear guidance to women regarding the risks and benefits of hormone therapy (Healy, 1995). In the early to mid-1990s, several large randomized, double masked, placebo-controlled clinical trials were initiated to address the limitations of previous observational studies. One clinical trial, the postmenopausal estrogen progestin (PEPI) study, looked at the effects of various hormone therapies on surrogate endpoints (e.g., lipid profiles) and found benefits on some endpoints (Writing Group for the PEPI Trial, 1995). Two larger clinical trials looked at the effects of HT on heart disease in women with pre-existing disease (i.e., a treatment as opposed to a prevention trial). Both of these found no benefit from therapy and possible risk in the immediate time frame around randomization (Grady et al., 2000; Herrington et al., 2000; Hulley et al., 1998). These data led the FDA to add a warning regarding the use of HT in the treatment of heart disease. Another randomized trial looked at the effects of HT in the treatment of dementia and found no effect (Henderson et al., 2000). Thus, data were emerging that suggested that HT, at best, had no effect in the treatment of heart disease or dementia.

However, the prevention question remained unanswered. Under the guidance and strong advocacy of the first woman director of the NIH, Dr. Bernadine Healy, the largest prospective longitudinal study ever undertaken to assess the primary causes of morbidity and mortality in older women was undertaken. The Women's Health Initiative (WHI) was begun in 1992 and enrolled 161,809 healthy postmenopausal women in an observational study and a set of randomized clinical trials: a low fat dietary intervention; calcium and vitamin D supplementation; and two trials of postmenopausal hormone use (i.e., women with and without a uterus were randomized to separate HT trials; The Women's Health Initiative Study Group, 1998). Approximately 27,000 women were enrolled into the hormone trials between 1993 and 1998 (16,608 women with a uterus; and 10,739 without a uterus). The primary outcome in the hormone trials was the development of coronary heart disease. Hip fracture was designated as a secondary outcome and invasive breast cancer was designated as a primary adverse outcome based on data from previous observational studies that demonstrated cancer risks. Secondary outcomes included examination of the impact of HT on other cardiovascular diseases, endometrial, colorectal and other cancers, and other types of fractures. In addition, ancillary studies evaluating the effects of HT on dementia, mild cognitive impairment, global cognitive functioning, and normal cognitive aging were added to the HT studies within the WHI (Shumaker et al., 1998).

In May of 2002, after a mean of 5.2 years of follow-up for women enrolled in the HT trials, the data and safety monitoring board for WHI recommended termination of the trial of estrogen plus progestin (brand name Prempro) versus placebo, citing, as the main factor, an increased risk of invasive breast cancer in the group receiving the combined HT therapy compared to placebo (Writing Group for the Women's Health Initiative Investigators, 2002). There was also an observed increase in cardiovascular events in women on active drug compared to those on placebo that began in the first year of treatment and persisted (Manson et al., 2003), as well as an increase in stroke risk (Wassertheil-Smoller et al., 2003), and no benefit for health-related quality of life (Hays et al., 2003). Finally, there was a doubling of risk of dementia for women in the combination hormone therapy group when compared to the placebo group (Shumaker et al., 2003) and no benefit from treatment for cognitive functioning (Rapp et al., 2003). These results had a major impact in the United States and worldwide as scientists, physicians, the pharmaceutical industry, and women began to grapple with the dissonance created by their beliefs about presumed benefits and the empirical data on actual risks (Bailar, 2003; "Estrogen and Progestin Use," 2003; Fletcher & Colditz, 2002; Grady, 2003; Grodstein, Clarkson, & Manson, 2003; Herrington & Howard, 2003; Hersh, Stefanick, & Stafford, 2004; Hovey, 2004; Ignelzi, 2003; Kolata, 2002; Letters to the Editor, 2002; Smith, 2003).

Further complicating this issue was the early termination of the estrogen-alone trial of the Women's Health Initiative on February 29, 2004 (The Women's Health Initiative Steering Committee, 2004). This trial included those women who had had a hysterectomy and thus were randomized to receive only estrogen therapy without a progestin agent (trade name Premarin) or to receive a placebo. The reason for the early termination of this trial was an increased risk of stroke in women randomized to the estrogen therapy group when compared to the placebo group. No significant adverse effects of estrogen-only therapy were observed for coronary heart disease, breast cancer, pulmonary embolism, or colorectal cancer. A benefit of estrogen therapy was noted for hip fracture in that women randomized to the estrogen therapy group had a lower incidence of hip fracture than those randomized to the placebo group. Data on the effects of Premarin on dementia and global cognitive functioning have been published recently, and indicated an increased risk of dementia and mild cognitive impairment, and a decline in global cognitive function in women on active therapy as opposed to placebo (Espeland et al., 2004; Shumaker et al., 2004).

Controversy Over the Research Findings

Since the WHI estrogen plus progestin clinical trial was stopped in 2002, there has been a barrage of opinions regarding the place of estrogen replacement therapy in women's health (Bailar, 2003; Fletcher & Colditz, 2002; Grady, 2003; Grodstein et al., 2003; Herrington & Howard, 2003; Letters to the Editor, 2002). Literally hundreds of articles and editorials have been published in the scientific and popular press discussing the value and validity of the WHI data. Prescription use of Premarin, Prempro, and other hormone therapies has plummeted to levels similar to those observed when they were prescribed for menopausal symptoms only—a use for which they are still recommended (Hensley & Landers, 2003; Mathews & Hensley, 2003).

With this kind of impact and in light of the long period of time during which it was believed that hormone therapy was beneficial, resistance to relinquishing belief in the benefits of HT has been strong. The WHI results have been roundly criticized on many points, including: WHI used only one HT drug regimen; the results are not applicable to women who go on short-term hormone therapy at the time of menopause for symptom relief; stopping the trials early reduced the precision of the estimates of the long-term effects of treatment on the primary and secondary outcomes; the absolute numbers of participants who suffered adverse events was very small, although the results were statistically significant; and women themselves will be unlikely to forego using HT, because it makes them feel better and improves their quality of life (Letters to the Editor, 2002).

As the final arbiter of drug safety in the United States, any move by the FDA to modify its position on HT could be viewed as a definitive judgment about the reliability and validity of the WHI results. After careful analysis of the WHI data as well as other data emerging in recent years, the FDA has reinforced and maintained its initial labeling for combination estrogen plus progestin HT (Prempro/Premphase Full Revised Label, 2003). HT is currently recommended for the treatment of peri- and postmenopausal symptoms. Further, women (and their physicians) are encouraged to use the lowest dose for the shortest period of time to address such symptoms. (It should be noted that there is no evidence, positive or negative, that a lower dose will lead to decreased adverse outcomes. However, in the absence of such evidence, the known immediate symptom-related benefits, and the low absolute risks of HT for women close to the menopausal years, the FDA determined this to be the most prudent labeling at this time.) HT is also recommended for the prevention and treatment of osteoporosis for those instances where other treatments are not effective and/or with the woman's full understanding of the risks of such use. These recommendations were extended to all combinations of estrogens and progestins. The FDA has also added a new warning to HT labeling (See Appendix) which, in addition to stating that HT should not be used for prevention of cardiovascular disease, includes information regarding the risks found by the WHI (Prempro/Premphase Full Revised Label, 2003). The same cardiovascular disease warning was also added to the Premarin drug label (Premarin Full Revised Label, 2003).

Clearly the FDA regards the WHI results as definitive, but one might wonder if the agency has erred in favor of caution. How should we as researchers and scientists regard the WHI results? The WHI represents the best and largest study on this issue to date. Randomization has removed the possibility that some selection process is driving the results as was likely in earlier observational studies and small, unmasked clinical trials. Placebo controls and double masking removed the possibility that bias was introduced inadvertently by data collectors, data analyzers, or the study participants. The size of the study, 16,000 (substantially larger than any study to date on HT), and the long follow-up period assures sufficient statistical power to detect clinically important differences even in small subsets of the larger sample. Consequently, when some argue that the sample was "too old" (mean age: 63), subset analyses on 1500 women under the age of 60 still provides sufficient power to detect clinically important as well as statistically significant differences in outcomes.

Still, the WHI is only one study and most researchers and policy analysts would agree that it is nonoptimal to base policy on one study even if the FDA is willing to do so. But we need not consider only this one study. There have been other clinical trials and nearly all are consistent with the findings of the WHI: higher risk of breast cancer, higher risk of blood clots (Grady et al., 2000; Hully et al., 1998), and no evidence of cardiovascular protection (Grady et al., 2000; Hodis et al., 2003; Hulley et al., 1998; Waters et al., 2002). The methods and the samples may have differed somewhat, limiting their generalizability, but their results are consistent with those of the WHI. This suggests that while researchers may attempt to fault certain aspects of the WHI study (e.g., sample is too old, length of time on HT was too long), it represents a strong evaluation of current practice and is consistent with previous findings regarding risks associated with HT.

Although the results of the WHI Estrogen + Progestin study and the WHI Estrogen Alone trials were statistically significant, the absolute numbers of participants who suffered adverse events were very small, leading to discussions that the results of these studies have been overblown and that they have alarmed women unnecessarily. The WHI Estrogen + Progestin trial estimated only 7 more coronary heart disease (CHD) events, 8 more strokes, 8 more pulmonary emboli, 8 more invasive breast cancers, and 21 more dementias for 10,000 person years. In the WHI Estrogen Alone Trial, 12 more strokes for 10,000 person years were found among those randomized to active hormone therapy. Although the numbers seem to be small, these figures need to be placed into context. It is estimated that the percentage of postmenopausal women using HT for some period is approximately 38% (Keating, 1999). Given the increasing numbers of women over the age of 50 years in the U.S. population, these estimates actually translate into high numbers of women who have the potential to be affected. The WHI investigators examined the net effect of these adverse events and, excluding the dementia risk, reported that the use of estrogen plus progestin therapy (Prempro) translated into 2 adverse events for 1,000 women treated for 1 year and 1 serious adverse event for 1 out of 100 women treated with Prempro for 5 years (Grady, 2003). These risks are actually very high when considered from the larger, public health perspective. Further, the WHI study was begun because physicians were placing asymptomatic menopausal and older (far past menopause) women on a drug for life with the hope it would decrease their risks for heart disease and dementia. Recent data estimate that the reduction in prescriptions of hormone therapy following the stopping of the WHI Estrogen plus Progestin Trial lead to a 65% reduction in adverse events or 8,000 cardiovascular, coronary heart disease, or breast cancer events related to the use of hormone therapy in 2003 (Hersh et al., 2004). Clearly prevention of coronary heart disease and cardiovascular diseases is not a recommended and appropriate use of these therapies in light of these observed outcomes.

In taking a second look at the value of the WHI (and other studies of hormone therapy), the implications for the health and well-being of older women, and the "lost promise" of hormone therapy, one has to wonder how these results could produce controversy. Specifically, in the face of such strong evidence that, at a minimum, hormone therapy does not help the postmenopausal woman beyond the osteoporosis benefit (which can be attained with alternative, lower-risk treatments) and can actually harm the aging woman, why has there been resistance among many physicians and women to changing beliefs and practice patterns? One approach to exploring this question is to examine the stakeholders in this debate: pharmaceutical manufacturers, postmenopausal women, and physicians. It is among these different constituencies that policy, practice, economics, and beliefs are at odds. For all stakeholders, the stakes are quite high.

Pharmaceutical Manufacturers

For hormone manufacturers, the stakes are billions of dollars in revenue from hormones and other drugs aimed at reducing symptoms and diseases of the aging process in women. For Wyeth of Madison, NJ, the maker of Premarin and Prempro, termination of the WHI hormone trials translated into dramatic declines in revenue and share price. According to the *The Wall Street Journal* (Hovey, 2004), after the WHI E+P trial was stopped midyear in July 2002, sales of Prempro and Premarin fell to $266 million in the first quarter of 2004. This represents a substantial decline in hormone-related revenues which were reported to be roughly $2.5 billion prior to the WHI study (Hensley, 2002). These declines in revenue paralleled a comparable drop in prescriptions for this therapy (Associated Press, 2004). Of course, the impact is not just reflected in sales figures and revenues. *Business Week* reported that Wyeth's stock fell sharply from $50 to $34.50 between July 2 and July 24, 2002, in response to the news from the WHI trial (Barrett, 2002). It had rebounded, trading at around $44 as of October 2003 (Market Information (WYE), 2003). However, by late March 2004, it had dropped to about $37 (Market Informtion (WYE), 2004), possibly in response to the first lawsuits filed in mid-February by plaintiffs based on claims from two diet drugs withdrawn from the market six years ago. These are expected to continue to burden the company's financial outlook (Hensley & Landers, 2003).

Wyeth is not the only pharmaceutical manufacturer at risk. Schering AG of Frankfurt, Germany, specializing in female fertility and hormone therapy, faced stock downgrades by market analysts shortly after the WHI study was halted ("Schering Defends Hormone Drugs," 2002). Nor are other pharmaceutical companies immune. No doubt feeling vulnerable, Noven Pharmaceuticals and Novartis, makers of estrogen patches, have argued that the WHI results may not extend to their products. Similarly, Novavax, Inc, is reported to have encountered problems from the FDA in bringing a new HT drug to market (Estrasorb). The firm expects, also, to face a diminished market for HT and difficulties in marketing their product because of the WHI study results (Barbaro, 2002). At the same time, not all pharmaceutical firms stand to lose. Eli Lilly and Merck, makers of Evista and Fosamax, two non-estrogen drugs aimed at reducing osteoporosis, may actually benefit from women's reluctance to use Prempro (Herper, 2002). Similarly, Pfizer (Lipitor) and Merck (Zocor) may benefit if women at risk for heart attack and cardiac death switch to these or other statins. In fact, an ironic "side effect" discovered in WHI, Heart and Estrogen/Progestin Replacement Study (HERS) and the Estrogen Replacement and Atherosclerosis (ERA)-Trial of the push for hormone therapy to prevent and treat heart disease was the underutilization by women and their physicians of drugs like statins that have known benefits with respect to heart disease (Grady et al., 2000).

In doing their job of making profits and serving their shareholders, pharmaceutical companies face many challenges. They appear to surmount these challenges very well. *The Economist* ("Fixing the Drugs Pipeline," 2004) reports that the pharmaceutical industry is "one of the biggest and most lucrative in the world with annual sales of around $400 billion" (p. 37). Drug company representatives will be the first to explain the very high research and development (R&D) costs that they face in the never-ending quest for market share. These R&D costs are often offered as justification for the very high prices charged for drugs U.S. consumers have come to rely on. Unfortunately, recent evidence suggests that while R&D costs are indeed high, marketing costs are rising, also, as drug companies struggle to maintain market share on a dwindling pool of moneymakers ("Fixing the Drugs Pipeline," 2004). Moreover, there is evidence that much of the R&D budget is spent, not on new and innovative drugs, but on bringing older drugs to market with only minor modifications, thus allowing manufacturers to extend patent periods and the higher prices associated with them (Hunt, 2002). There is also some evidence that this strategy may backfire as increasing numbers of firms face expiring patents for major products, competition from generics and over the counter versions of former prescription drugs, and a shortfall of new and innovative "blockbusters" ("Fixing the Drugs Pipeline," 2004; Hensley & Landers, 2003; Landers, 2004). The dearth of innovation has been attributed to several causes among which are the industry's main interest in drugs that produce large annual revenue ($1 billion or more), increasingly complex disease processes that must be mastered in order to be treated successfully, and large investments made a decade ago in new technologies aimed at discovering new drugs that have not paid off as expected and may have actually backfired ("Fixing the Drugs Pipeline," 2004; Landers, 2004).

Underlying all these business challenges is a very real and less tractable challenge that derives from the nature of the pharmaceutical business: the randomness and uncertainty of many diseases. From a purely business point of view, the most reliable profits can be made from disease processes and conditions that have high incidence and prevalence, where future demand can be predicted with a high degree of certainty and is not dependent on the random occurrence of disease processes. If the condition is chronic, even more certainty about market demand is guaranteed. A population like that of the United States, where a large number of baby boomers are advancing steadily into old age, provides the creative marketer with ample opportunity to plumb untapped reservoirs of demand if only the potential demanders can be educated to recognize their condition and the treatments available for it (Moynihan, 2003; Moynihan, Heath, & Henry, 2002). Clearly, menopause and the symptoms associated with it provide drug companies with such an opportunity. It is also possible that the ensuing controversy about the risks of hormone therapy may enhance this opportunity.

Physicians

For physicians, what appears to be at stake is that most precious commodity, the commodity which Abraham Flexner and the American Medical

Association (AMA) struggled to create and maintain, the commodity without which the profession would not enjoy its well-earned reputation for good science and ethical medicine—professional credibility (Starr, 1982). This group of professionals, sworn to "do no harm," is now confronted with the results of one large, prospective, well-designed study that the FDA believed provided sufficient evidence to warrant modifying the labeling of hormone therapy products—a study that is also consistent with the results of several other well-designed studies. These results appear to indicate that some harm may have been done and that they are contrary to the practice patterns of many physicians who have strongly advocated for the use of hormone therapy among postmenopausal women ("Estrogen and Progestin Use," 2003). Thus, some physicians are confronted with a state of cognitive dissonance over these recent study results.

Coupled with this dissonance is selection bias with respect to the women seen and treated by health care providers. That is, women who seek care during the menopause are those women who are most likely to be experiencing moderate to severe symptoms associated with menopause, are better educated, and have better access to health care services (Grodstein & Stampfer, 1995). Physicians have learned that treating these symptoms with hormone therapy does, in fact, provide relief to most of their patients. How can a drug that does such immediate good and that appeared to have the potential to do such long-term good, be so wrong? And, how does an informed physician guide a woman through what has evolved as a complex risk/benefit analysis?

Further confounding the issue are those physicians who receive money from drug companies for research, foundations, and endorsements, and who may or may not disclose these conflicts of interest when they speak publicly on medical matters related to those drug companies or their products. These conflicts of interest are usually not reported in the popular press, while medical journals have only recently begun to require such reporting (Davidoff et al., 2001).

There can be no doubt that partnerships between academic researchers and industry lead to greater technological advances and improved access to new drugs and medical devices for patients. However, much of the early history of the medical profession is dominated by efforts to build and maintain credibility. Indeed, many of these efforts focused on driving a wedge between physicians and drug manufacturers in an effort to remove such conflicts of interest (Hilts, 2003; Starr, 1982). Failure to adhere to codes of conduct covering conflicts of interest, both in research journals and in the popular press, and failure to adhere to gold-standard research methodology in shaping standards of care are threats to this hard-won credibility and do great harm, both to the profession and to the patients it serves. If financial interests influence physician's practice patterns and beliefs about drug efficacy, even when faced with results from sound studies that indicate the absence of benefit, or worse, the presence of harm, then, not only will credibility be lost, but controversy is sure to ensue.

Finally, it may be that one of the biggest threats to this credibility is simply the large and ever-increasing body of research that informs medical practice and the very real human limitations in assimilating and evaluating it. However, without formal consensus grounded in sound research and translated into standards of care, physicians' individual beliefs and practice patterns will continue to shape medical care protocols and to fuel controversy. The absence of a formal mechanism for such consensus and translation may contribute to the HT controversy.

Women

For peri- and postmenopausal women, the stakes regarding the use of hormone therapy are personal. It has been estimated that as many as two-thirds of menopausal women in the United States will experience vasomotor symptoms, such as hot flushes and night sweats (Achilles & Leppert, 1997). For some women, these symptoms will be of mild to moderate levels of discomfort, but others will experience severe menopausal symptoms. The lack of safe and effective alternatives for the treatment of these symptoms is a genuine health problem. While some criticisms of the WHI results have focused on the relatively small absolute risk of an adverse event, clearly the negative long-term sequelae of a hormone therapy-associated breast cancer, cardiovascular disease, and dementia are more severe for those women who develop these conditions than the often inconvenient and unpleasant, but usually short-lived symptoms for which relief is proximately sought.

In conjunction with the symptoms of menopause, however, are beliefs surrounding the menopause and its meaning for a woman's post-reproductive life. When hormone therapy was marketed in the 1950s and 1960s, it was presented as a way for women to maintain or regain their femininity and a more youthful appearance, and be a vital member of society (Healy, 1995). There is significant evidence (both anecdotal and from news reports) suggesting that many women still believe that HT represents that most elusive and alluring of all states of being—eternal youth or, at least, the prolongation of youth. Kolata (2002) quotes one gynecologist who provides anecdotal evidence that women are reluctant to give up HT because of a belief that HT promotes a youthful appearance. Other news reports document similar attitudes (Ignelzi, 2003; Smith, 2003). Going beyond anecdotal reports, larger survey studies are needed to determine the degree to which these beliefs among women are maintained post-WHI. The positive effect of HT on decreasing vaginal dryness and atrophy, which tend to occur after menopause, has also made some women reluctant to stop using hormone therapy (Achilles & Leppert, 1997). Given the added emphasis on youth in our society and that aging is not honored for men or women (arguably less so for women), it is understandable that methods to prevent or delay the aging process are sought. HT promised to address this need in women—prolonged youth and sensuality—while simultaneously providing benefits against heart disease, osteoporosis, and dementia. Unfortunately, the results from WHI and other trials have not demonstrated any of these anticipated effects.

Discussion

The controversy regarding the use of hormone therapy in peri- and postmenopausal women is expected to continue for some time. With the early termination of the WHI Estrogen + Progestin Trial and the very recent termination of the WHI Estrogen Alone Trial, as well as results from other related studies, research and discussion about hormone therapy for women remain dominant topics among women's health researchers, physicians, regulators, and women themselves. The competing interests of the various constituencies, most notably pharmaceutical companies, physicians, and aging women, fuel the discussions and tend to produce controversy. The stakes for each group are distinct and varied. For large pharmaceutical companies, currently facing a future of lower profits from a dwindling pool of moneymaking drugs and higher spending on marketing to maintain market share to offset some of this profit loss, billions of dollars in anticipated revenue are at stake along with pressures to serve company shareholders better. Physicians, who oftentimes have little training in scientific research methods necessary to accurately evaluate research findings and who face the daunting task of remaining current in a field in which information grows and changes at increasing rates, may rely on drug company representatives for information on disease treatment. Aging women themselves, faced with few effective alternatives for the treatment of bothersome and severe menopausal symptoms and led for many years to believe that other effects of aging could be held at bay with HT, are understandably confused and reluctant to relinquish anything that relieves these symptoms or that provides even the illusion that aging can be slowed. All these competing and complementary forces fuel the controversy.

After reviewing the accumulated evidence regarding hormone therapy and its potential uses for the primary and secondary prevention of chronic diseases in women, it is clear that risks of long-term use of hormone therapy outweigh benefits. The question of the use of hormone therapies for menopausal symptom relief, its primary FDA approved use, is more difficult. Although hormone therapy has been shown to be a very effective treatment for menopausal symptoms, women who use this therapy for symptom relief are placing themselves at a small—but real—increased risk for coronary and cardiovascular events, emboli, cognitive decline, and breast cancer. Though small, the risks are not insignificant and each woman and her physician must weigh any anticipated benefits of taking hormone therapy against the possible risks. Further, although substantial research is underway currently, at this time investigators cannot identify which women are at higher risk for the adverse outcomes associated with HT. In the absence of other clearly effective treatments for menopausal symptoms, this leaves women with severe symptoms in a quandary. It is expected that research will continue on other preparations of hormone therapies than the ones chiefly discussed in this article, as well as on other hormone delivery systems, such as transdermal estrogen patches and nasal sprays. Both may hold promise in more safely treating symptoms in peri-menopausal and postmenopausal women. Until that time, controversy regarding hormone therapy and its place in women's health is likely to continue.

References

Achilles, C., & Leppert, P. C. (1997). The menopausal woman. In P. C. Leppert & F. M. Howard (Eds.), *Primary care for women* (pp. 97–102). Philadelphia: Lippincott-Raven Publishers.

Amundsen, D. W., & Diers, C. J. (1973). Age of menopause in medieval Europe. *Human Biology, 25,* 605.

Associated Press. (2004, January 6). Hormone prescriptions drop after risks found in study. *The Wall Street Journal.* Retrieved January 29, 2004, from http://www.WSJ.com

Avis, N. E., Crawford, S., Stellato, R., & Longcope, C. (2001). Longitudinal study of hormone levels and depression among women transitioning through menopause. *Climacteric, 4,* 243–249.

Bailar, J. (2003). Hormone-Replacement therapy and cardiovascular diseases. *The New England Journal of Medicine, 349*(6), 521–522.

Barbaro, M. (2002, September 17). Novavax faces tough sales job on drug: Hormone-replacement concerns could sink Estrasorb even if FDA doesn't. *The Washington Post* (final ed.). Retrieved April 24, 2004, from Lexis-Nexis database.

Barnabei, V. M., Grady, D., Stovall, D. W., Cauley, J. A., Lin, F., Stuenkel, C. A., et al. (2002). Menopausal symptoms in older women and the effects of treatment with hormone therapy. *Obstetrics and Gynecology, 100,* 1209–1218.

Barrett, A. (2002, July 25). Good news about HRT—for investors. *BusinessWeek.* Retrieved October 13, 2003, from http://www.businessweek.com/technology/content/jul2002/tc20020725_9019.htm

Benedek, T. (1950). Climacterium: A developmental phase. *Psychoanalysis Quarterly, 19,* 1–27.

Birge, S. M. (1997). The role of estrogen in the treatment and prevention of dementia: Introduction. *American Journal of Medicine, 103*(suppl.), 1S–2S.

Buxton, C. L., & Engle, E. T. (1939). Effects of the therapeutic use of diethylstilboestrol. *Journal of the American Medical Association, 113,* 2318–2320.

Davidoff, F., DeAngelis, C. D., Drazen, J. M., Hoey, J., Hojgaard, L., Horton, R., et al. (2001). Sponsorship, authorship and accountability. *Lancet, 358,* 854–856.

Deutsch, H. (1945). Psychology of women: A psychoanalytic interpretation. In (Ed.), *Motherhood: Vol. 11* (pp. 456–491). New York: Grune and Stratton.

Espeland, M. A., Rapp, S. R., Shumaker, S. A., Brunner, R., Manson, J. E., Sherwin, B. B., et al. (2004). Conjugated equine estrogens and global cognitive function in postmenopausal women. *JAMA, 291,* 2959–2968.

Estrogen and progestin use in peri- and post-menopausal women. (2003, September). Position Statement of the North American Menopause Society. *Menopause, 10*(6), 497–506.

Fixing the drugs pipeline. (2004). *The Economist* (March 13th–19th), 37–38.

Fletcher, S. W., & Colditz, G. A. (2002). Failure of estrogen plus progestin therapy for prevention. *Journal of the American Medical Association, 288*(3), 366–368.

Friedman, L. M., Furberg, C. D., & DeMets, D. L. (1996). *Fundamentals of clinical trials* (third ed.). St. Louis, MO: Mosby Press.

Gannon, L. (1990). Endocrinology of menopause. In R. Formanek (Ed.), *The meanings of menopause: Historical, medical and clinical perspectives* (pp. 179–237). Hillsdale, NJ: Analytic Press.

Grady, D. (2003). Postmenopausal hormones—Therapy for symptoms only. *The New England Journal of Medicine, 348*(19), 1835–1837.

Grady, D., Rubine, S. M., Petitti, D. B., Fox, C. S., Black, D., Ettinger, B., et al. (1992). Hormone therapy to prevent disease and prolong life in postmenopausal women. *Annals of Internal Medicine, 117*, 1016–1037.

Grady, D., Wenger, N. K., Herrington, D., Khan, S., Furberg, C., Hunninghake, D., et al. (2000). Postmenopausal hormone therapy increases risk for venous thromboembolic disease: The heart estrogen/progestin replacement study. *Annals of Internal Medicine, 132*(9), 689–696.

Greendale, G., Reboussin, B., Hogan, P., Barnabei, V. M., Shumaker, S., Johnson, S., et al. (1998). Symptom relief and side effects of postmenopausal hormones: Results from the postmenopausal estrogen/progestin interventions trial. *Obstetrics and Gynecology, 92*, 982–988.

Grodstein, F., Clarkson, T. B., & Manson, J. E. (2003). Understanding the divergent data on post-menopausal hormone therapy. *The New England Journal of Medicine, 348*(7), 645–650.

Grodstein, F., & Stampfer, M. (1995). The epidemiology of coronary heart disease and estrogen replacement in postmenopausal women. *Progress in Cardiovascular Diseases, 38*(3), 199–210.

Herper, M. (2002, September 4). Wyeth cautious on long-term hormone use. *Forbes*. Retrieved October 13, 2003, from http://www.forbes.com/2002/09/04/0904wye.html

Hays, J., Ockene, J. K., Brunner, R. L., Kotchen, J. M., Manson, J. E., Patterson, R. E., et al. (2003). Effects of estrogen plus progestin on health-related quality of life. *The New England Journal of Medicine, 348*(19), 1839–1854.

Healy, B. (1995). Menopause: The new prime time. In *A new prescription for women's health* (pp. 173–213). New York: Viking Penguin.

Henderson, V. W., Paganini-Hill, A., Miller, B. L., Elble, R. J., Reyes, P. F., Shoupe, D., et al. (2000). Estrogen for Alzheimer Disease in women: Randomized, double-blind, placebo-controlled trial. *Neurology, 54*, 295–301.

Hensley, S. (2002, July 10). Wyeth stock slides as study casts shadow on medicines. *The Wall Street Journal*. Retrieved October 23, 2003, from http://www.WSJ.com

Hensley, S., & Landers, P. (2003). Drug companies report pain. *The Wall Street Journal Online*, October 23, 2003. Retrieved October 23, 2003 from, www.WSJ.com.

Herrington, D. M., & Howard, T. D. (2003). From presumed benefit to potential harm—Hormone therapy and heart disease. *The New England Journal of Medicine, 349*(6), 519–521.

Herrington, D. M., Reboussin, D. M., Broshnihan, K. B., Sharp, P. C., Shumaker, S. A., Snyder, T. E., et al. (2000). Effects of estrogen replacement on the progression of coronary artery atherosclerosis. *The New England Journal of Medicine, 343*, 522–529.

Hersh, L., Stefanick, M. L., & Stafford, R. S. (2004). National use of postmenopausal hormone therapy. Annual trends and response to recent evidence. *Journal of the American Medical Association, 291*, 47–53.

Hilts, P. J. (2003). *Protecting America's health: The FDA, business, and one hundred years of regulation*. New York: Alfred J. Knopf.

Hodis, H. N., Mack, W. J., Azen, S. P., Lobo, R. A., Shoupe, D., Mahrer, P. R., et al. (2003). Hormone therapy and the progression of coronary-artery atherosclerosis in postmenopausal women. *The New England Journal of Medicine, 349*, 535–545.

Hovey, H. H. (2004, April 22). Wyeth's profit declines by 41% skewed by big year-earlier gain. *The Wall Street Journal*. Retrieved April 22, 2004, from http://www.WSJ.com

Hulley, S., Grady, D., Bush, T., Furberg, C., Herrington, D., Riggs, B., et al. (1998). Randomized trial of estrogen plus progestin for secondary prevention of coronary heart disease in post-menopausal women. *Journal of the American Medical Association, 280*, 605–613.

Hunt, M. (2002, May). *Changing patterns of pharmaceutical innovation*. Washington, DC: National Institute for Health Care Management.

Ignelzi, R. J. (2003, November 24). Women weigh the risks vs. the benefits of hormone therapy. *Copley News Service*. Retrieved April 24, 2004, from Lexis-Nexis database.

Keating, N. L., Cleary, P. D., Rossi, A. S., Zaslavsky, A. M., & Ayanian, J. Z. (1999). Use of hormone replacement therapy by postmenopausal women in the United States. *Annals of Internal Medicine, 130*, 545–553.

Kolata, G. (2002, November 10). Menopause without pills: Rethinking hot flashes. *The New York Times* (Sunday, late ed.-final). Retrieved April 22, 2004, from Lexis-Nexis database.

Kreling, D., Mott, D., Widerholt, J., Lundy, J., & Levitt, L. (2001). *Prescription drug trends: A chartbook update*. Menlo Park, CA: Kaiser Family Foundation.

Landers, P. (2004, February 24). Drug industry's big push into technology falls short. *The Wall Street Journal*. Retrieved February 24, 2004, from http://www.WSJ.com

Letters to the Editor. (2002). *Journal of the American Medical Association, 288*(22), 2819–2825.

MacBryde, C. M., Freedman, H., & Loeffel, E. (1939). Studies on stilbestrol: Preliminary statement. *Journal of the American Medical Association, 113*, 2320.

MacLennan, A., Lester, S., & Moore, V. (2001). Oral estrogen replacement therapy versus placebo for hot flushes: A systematic review. *Climacteric, 4*, 58–74.

Manson, J. E., Hsia, J., Johnson, K. C., Rossouw, J. E., Assaf, A. R., Lasser, N. L., et al. (2003). Estrogen plus progestin and the risk of coronary heart disease. *The New England Journal of Medicine, 349*(6), 523–534.

Market Information (WYE). (2003). *New York Stock Exchange*. Retrieved October 27, 2003, from http://www.nyse.com/marketinfo

Market Information (WYE). (2004). *New York Stock Exchange*. Retrieved March 23, 2004, from http://www.nyse.com/marketinfo

Mathews, A., & Hensley, S. (2003, October 8). Hormone-Therapy debate grows. *The Wall Street Journal*. Retrieved October 13, 2003, from http://www.wsj.com

McKinlay, S., Brambilla, D., & Posner, J. (1992). The normal menopause transition. *Maturitas, 14*, 103–115.

Moynihan, R. (2003). The making of a disease: Female sexual dysfunction. *British Medical Journal, 326*, 45–47.

Moynihan, R., Heath, I., & Henry, D. (2002). Selling sickness: The pharmaceutical industry and disease mongering. *British Medical Journal, 324*, 886–889.

National Center for Health Statistics. (1999). *Health, United States, 1999 with the health and aging chartbook* (PHS 99-1232). Hyattsville, MD: U.S. Department of Health and Human Services.

Nelson, H. D., Humphrey, L. L., Nygren, P., Teutsch, S. M., & Allan, J. D. (2002). Postmenopausal hormone replacement: Scientific review. *Journal of the American Medical Association, 288*, 872–881.

Premarin full revised label. (2003, January). Retrieved December 2, 2004, from: http://www.fda.gov/medwatch/SAFETY/2003/premarin_PI.pdf

Prempro/Premphase full revised label. (2003, January). Retrieved June 23, 2004, from: http://www.fda.gov/medwatch/SAFETY/2003/prempro_PI.pdf

Rapp, S. R., Espeland, M. A., Shumaker, S. A., Henderson, V. W., Brunner, R. L., Manson, J. E., et al. (2003). Effect of estrogen plus progestin on global cognitive function in postmenopausal women: The Women's Health Initiative Memory Study: A randomized controlled trial. *Journal of the American Medical Association, 289*(20), 2663–2672.

Resnick, S. M., Maki, P. M., Golski, S., Kraut, M. A., & Zonderman, A. B. (1998). Effects of estrogen replacement therapy on PET cerebral blood flow and neuropsychological performance. *Hormones and Behavior, 34*, 171–182.

Reuben, D. (1969). *Everything you always wanted to know about sex*. New York: Harper Collins.

Robinson, R. W., Cohen, W. D., & Higano, N. (1958). Estrogen replacement therapy in women with coronary atherosclerosis. *Annals of Internal Medicine, 48*, 95–101.

Schmidt, P. J., & Rubinow, D. (1991). Menopause-Related affective disorders: A justification for further study. *American Journal of Psychiatry, 148*, 844–852.

Seaman, B. (2003). *The greatest experiment ever performed on women*. New York: Hyperion Books.

Schering defends hormone drugs in the wake of halt of U.S. study. (2002, July 10). *The Wall Street Journal*. Retrieved October 13, 2003, from http://www.wsj.com

Shorr, E., Robinson, F. H., & Papanicolaou, G. N. (1939). A clinical study of the synthetic estrogen, stilbestrol. *Journal of the American Medical Association, 113*, 2312–2318.

Shumaker, S. A., Legault, C., Rapp, S. R., Thal, L., Wallace, R. B., Ockene, J. K., et al. (2003). Estrogen plus progestin and the incidence of dementia and mild cognitive impairment in postmenopausal women: The women's health initiative memory study: A randomized controlled trial. *Journal of the American Medical Association, 289*(20), 2651–2662.

Shumaker, S. A., Legault, C., Kuller, L., Rapp, S. R., Thal, L., Lane, D. S., et al. (2004). Conjugated equine estrogens and incidence of probable dementia and mild cognitive impairment in postmenopausal women. *JAMA, 291*, 2947–2958.

Shumaker, S. A., Reboussin, B. A., Espeland, M. A., Rapp, S. R., McBee, W. L., Dailey, M., et al. (1998). The Women's Health Initiative Memory Study (WHIMS): A trial of the effect of estrogen therapy in preventing and slowing the progression of dementia. *Controlled Clinical Trials, 19*, 604–621.

Smith, S. (2003, July 20). Hormone therapy's rise and fall: Science lost its way and women lost out. *The Boston Globe* (Sunday, 3rd ed.), p. A1.

Spake, A. (2002, November 18). The menopausal marketplace. *U.S. News & World Report*, pp. 1–6. Retrieved July 2, 2004, from http://USNews.com

Stampfer, M. J., & Colditz, G. A. (1991). Estrogen replacement therapy and coronary heart disease: A quantitative assessment of the epidemiologic evidence. *Preventive Medicine, 20*, 47–63.

Starr, P. (1982). *The social transformation of American medicine.* New York: Basic Books.

Swartzman, L. D., & Leiblum, S. R. (1987). Changing perspectives on the menopause. *Journal of Psychosomatic Obstetrics and Gynecology, 6*, 11–24.

U.S. Census Projections. (2003). Retrieved, from U.S. Census Web site: http://www.census.gov/population/projections/nation/summary/np-13-b.pdf

Utian, W. H. (1980). *Menopause in modern perspective: A guide to clinical practice.* New York: Appleton-Century-Crofts.

Wassertheil-Smoller, S., Hendrix, S. L., Limacher, M., Heiss, G., Kooperberg, C., Baird, A., et al. (2003). Effect of estrogen plus progestin on stroke in postmenopausal women. *Journal of the American Medical Association, 289*, 2673–2684.

Waters, D. D., Alderman, E. L., Hsia, J., Howard, B. V., Cobb, F. R., Rogers, W. J., et al. (2002). Effects of hormone replacement therapy and antioxidant vitamin supplements on coronary atherosclerosis in postmenopausal women: A randomized controlled rrial. *Journal of the American Medical Association, 288*, 2432–2440.

Wilson, R. A. (1966). *Feminine forever.* New York: M. Evans and Company, Inc.

The Women's Health Initiative Steering Committee. (2004). Effects of conjugated equine estrogen in postmenopausal women with hysterectomy. *Journal of the American Medical Association, 291*, 1701–1712.

The Women's Health Initiative Study Group. (1998). Design of the Women's Health Initiative Clinical Trial and Observational Study. *Controlled Clinical Trials, 19*, 61–109.

Writing Group for the PEPI Trial. (1995). Effects of estrogen or estrogen/progestin regimens on heart disease risk factors in postmenopausal women: The Postmenopausal Estrogen/Progestin Interventions (PEPI) Trial. *Journal of the American Medical Association, 273*(3), 199–208.

Writing Group for the Women's Health Initiative Investigators. (2002). Risks and benefits of estrogen plus progestin in healthy postmenopausal women. *Journal of the American Medical Association, 288*(3), 321–333.

Yaffe, K., Sawaya, G., Lieburg, I., & Grady, D. (1998). Estrogen therapy in postmenopausal women: Effects on cognitive function and dementia. *Journal of the American Medical Association, 285*, 1489–1499.

MICHELLE J. NAUGHTON, Ph.D. is Associate Professor in the Department of Public Health Sciences at the Wake Forest University School of Medicine in Winston-Salem, North Carolina. She completed her PhD in Sociology from the University of Iowa, and a post-doctoral fellowship and MPH in Epidemiology at the University of Minnesota. Her research interests and funded projects are in the areas of health-related quality of life, outcomes research, cancer survivorship, and women's health issues. She is also a Co-Investigator of the Clinical Facilitation Center for the Women's Health Initiative.

ALISON SNOW JONES, Ph.D. is a health economist and Assistant Professor in the Department of Public Health Sciences, Section on Social Sciences and Health

Policy, at the Wake Forest University School of Medicine. Her research focuses on women's health issues related to domestic violence and alcohol abuse, and on methodological issues pertaining to health effects of self-selected risk behaviors.

SALLY A. SHUMAKER, Ph.D., is Full Professor in the Departments of Public Health Sciences and Internal Medicine, Founding Director of the National Women's Health Center of Excellence, Director of the Office of Intercampus and Community Program Development, and Associate Dean for Research at the Wake Forest University School of Medicine. Dr. Shumaker earned her PhD in social psychology at the University of Michigan, and has published extensively in the areas of health-related quality of life, women's health and aging, the relationship between hormone therapy and cognition and cerebral changes, and related topics. She has served as Principal Investigator and Co-investigator on a number of government and industry sponsored grants, including the Women's Health Initiative Memory Study, the Clinical Facilitation Center for the Women's Health Initiative Study, and the Claude D. Pepper Center on Aging.

Appendix: New FDA Warning for HT based on WHI Study Results

Estrogens and progestins should not be used for the prevention of cardiovascular disease.

The Women's Health Initiative (WHI) study reported increased risks of myocardial infarction, stroke, invasive breast cancer, pulmonary emboli, and deep vein thrombosis in postmenopausal women during 5 years of treatment with conjugated equine estrogens (0.625 mg) combined with medroxyprogesterone acetate (2.5 mg) relative to placebo (see **CLINICAL PHARMACOLOGY, Clinical Studies**) Other doses of conjugated estrogens and medroxyprogesterone acetate, and other combinations of estrogens and progestins were not studied in the WHI and, in the absence of comparable data, these risks should be assumed to be similar. Because of these risks, estrogens and progestins should be prescribed at the lowest effective doses and for the shortest duration consistent with treatment goals and risks for the individual woman.

PREMPRO or PREMPHASE therapy is indicated in women who have a uterus for the following:

1. Treatment of moderate to severe vasomotor symptoms associated with the menopause.
2. Treatment of moderate to severe symptoms of vulvar and vaginal atrophy associated with the menopause. When prescribing solely for the treatment of symptoms of vulvar and vaginal atrophy, topical vaginal products should be considered.
3. Prevention of postmenopausal osteoporosis. When prescribing solely for the prevention of postmenopausal osteoporosis, therapy should only be considered for women at significant risk of osteoporosis and non-estrogen medications should be carefully considered.

From "Prempro/Premphase Full Revised Label," January, 2003. Retrieved December 2, 2004, from *http://www.fda.gov/medwatch/SAFETY/2003/prempro_PI.pdf*

Journal of Social Issues, Vol. 61, No. 1, 2005, pp. 181–191

Controlling Birth: Science, Politics, and Public Policy

Nancy Felipe Russo* and **Jean E. Denious**

Arizona State University

Reproductive technologies raise a host of social and legal issues that challenge basic values and create intense controversy. If researchers wish to inform public policies related to reproductive technologies, they must understand how the scientific enterprise is being manipulated and research findings are being misrepresented to justify a particular social agenda and restrict access to contraception and abortion. To counter these distortions, scientists must defend the science advisory process, be involved in dissemination of their research findings beyond simple publication in scientific journals, and actively work to ensure that the findings are not misrepresented to the public.

Over 30 years ago, Shulamith Firestone (1970) identified "freeing of women from the tyranny of their reproductive biology by every means available . . ." as a necessary condition for the equality of women in society, asserting "already we have a (hard won) acceptance of 'family planning,' if not contraception for its own sake" (p. 206). From today's vantage point, however, applications of reproductive technologies are mired in controversy and their implications for advancing women's status, health, and well-being are unclear—a reminder that technologies are but tools that serve the goals and motives of the powers that be (Beckman & Harvey, this issue).

As Linda Beckman and Marie Harvey (this issue) observe, "reproductive technologies" cover a broad terrain, encompassing techniques designed to prevent unwanted pregnancies and births to those aimed at enabling couples to conceive and bear healthy children. Whatever their specific purpose, these technologies share complex ethical, moral, and social issues, including those related to equality, women's empowerment, racism, poverty, and the role of science in policy making.

---

*Correspondence concerning this article should be addressed to Nancy Felipe Russo, Department of Psychology, Box 871104, Arizona State University, Tempe, AZ 85287-1104 [e-mail: nancy.russo@asu.edu].

We would like to thank Heather Boonstra and Allen Meyer for their comments on the manuscript.

In particular, modern reproductive technologies have brought historically private events into public view, such that basic beliefs about the family, parenthood, gender roles, and relationships are threatened and issues politicized (see Ciccarelli & Beckman, this issue).

The Political Context

Technologies designed to control birth, i.e., contraception and abortion, have been especially controversial (Russo & Denious, 1998), resulting in policies and politics becoming more closely intertwined. Indeed, birth control technologies have received public policy debate the longest of the modern reproductive technologies (Gordon, 1990). Consequently, a highly mobilized, politically sophisticated, and well-funded pro-life movement that advocates legal personhood be conferred at conception has emerged and flourished (Solinger, 1998). The political context is now such that reasoned discussions of reproductive policies and practices are distorted by the influence of that movement. The issues and themes found in debates about abortion are thus echoed in concerns raised about new forms of birth control such as emergency contraception (see Sherman, this issue).

Public policies operate at county, state, federal, and international levels, and originate in legislative, judicial, and executive branches of government (Lindblom & Woodhouse, 1993). The resulting, often inconsistent, patchwork of laws and policies makes it difficult for women to obtain a clear picture of their options, rights, and responsibilities with respect to the use of any particular technology. Even with *Roe v. Wade* (1973) some women, particularly young women, still must travel to different states to obtain an abortion. The politicization of the debate on abortion is further complicated by the use of restrictive policies on infertility treatments that involve embryo reduction as a stalking horse for the ultimate goal of restricting abortion rights. As Ciccarelli and Ciccarelli (this issue) note, the situation with regard to infertility treatments is particularly complex, with some jurisdictions covered by legislation, and others bound by court rulings where legislatures have failed to take action.

New technologies create options—and force decisions—in areas traditionally considered determined by biological destiny and/or "God's will," bringing religious interests to the policy table. Similarly, with the medicalization of reproductive-related events—from menstruation, to pregnancy and birth, to menopause—comes commercial interests (Stanworth, 1987). Business has a powerful influence on policymaking, and corporate involvement can complicate effective regulation against questionable business practices. Medicalization of reproductive events also gives health rationales increased weight in policy decision making (Lee, 2003).

In this context, building consensus among diverse players on policy goals becomes a necessary condition for positive change to occur. The complexities of these efforts are compounded by cultural differences as well as economic dis-parities that influence access, acceptability, and efficacy of various technologies (Severy & Newcomer, this issue). In such a politicized context, it is important to remember that policy making is ultimately a political process, and corporations and non-governmental organizations (including women's organizations, medical societies, and religious institutions) are key political players. If researchers seek to inform that process, their intended audience must extend beyond policy makers and include stakeholders as well as the public at large (Lindblom & Woodhouse, 1993).

A Moralistic Agenda

Although abortion is a mobilizing issue for social conservatives, they have a broader agenda reflected in public debates on a variety of health issues related to sexuality and reproduction. This agenda involves promotion of abstinence, moral-istic condemnation of individuals whose sexual behavior is not confined to hetero-sexual marriage, and erosion of the separation of church versus state as provided in the U.S. Constitution. It reflects, also, a "fundamental hostility toward contra-ceptive service programs for pregnancy prevention and promotion of condom use to prevent HIV/AIDS" (Cohen, 2002, p. 2).

Recent legislation articulating the U.S. approach for addressing the global AIDS problem demonstrates the extent to which socially conservative ideology has shaped public policy on health. Pro-life advocates have lobbied vigorously for defunding organizations that provide abortion (e.g., family planning clinics), and have pushed for ways to extend the "global gag rule," which requires that organizations abstain from abortion-related activities as a condition for receiv-ing family planning funds from the U.S. Agency for International Development (Holloway, 2001). When attempts to extend the gag rule to HIV/AIDS funding were unsuccessful, the Pro-life Caucus in the U.S. House of Representatives de-manded concessions for the bill's passage. Among them was making promotion of abstinence a priority in U.S. HIV prevention programs. In addition to amending the bill to reserve at least a third of prevention funds for "abstinence-until-marriage" programs, the interests of faith-based programs were protected against having to "endorse, utilize, or participate in a prevention method or treatment program" against their religious or moral objection (Boonstra, 2003b, p. 3).

This moralistic agenda is reflected, also, in the distortion of the science ad-visory process through the appointment (and removal) of advisers based on their agreement with a conservative social agenda rather than their scientific expertise. The dismissal of basic scientists from the President's Council is but one example (Weiss, 2004). Another is the appointment of W. David Hagar to the FDA's Re-productive Health Drugs Advisory Committee. A physician known for refusing to prescribe contraceptives to unmarried women, he has a documented record of "making medical choices influenced by personal religious beliefs rather than based on scientific research or clinical experience . . ." (Boonstra, 2003, p. 2).

Corruption of the Science Advisory Process

Both appointments to scientific advisory panels and public health information activities of the Department of Health and Human Services (DHHS) have traditionally been considered insulated from politics. The legislation that established the system of advisory committees to ensure governmental access to expert advice and diverse opinions permits the president or the appointing agency substantial leeway in making committee appointments. It requires only that membership be fairly balanced and not influenced by any special interest (including the appointing authority). The appointment of individuals outside the scientific mainstream, however, led the American Association for the Advancement of Science to pass a resolution emphasizing that "selection, removal, or replacement of committee members 'based on criteria extraneous either to the scientific, technical, or medical issues . . . compromises the integrity of the process of receiving advice and is inappropriate'" (Boonstra, 2003a, p. 1).

Central to the science advisory process is the system of peer review that functions as a quality control mechanism to uphold the highest possible standards for scientific research. Although by no means a perfect process, it has become the minimum standard for credibility of scientific contribution. Research funding, scientific publications, and even the promotion and tenure of academic scientists, are determined by the judgments of one's scientific peers (Drazen & Ingelfinger, 2003). Of most concern here is the unprecedented interference with the peer review processes used by scientific bodies at National Institutes of Health (NIH) and other research agencies (Leshner, 2004; Reppert, 2004). As described below, this interference includes the appointment of individuals without appropriate scientific credentials to oversee scientific decision making and to participate in committees that determine scientific merit of research proposals. There are also attempts to deny funding to specific grants (already approved through normal channels of the National Institutes of Health) that address sexual health and behavior, such as research on HIV/AIDs and risk of sexually transmitted infections (STIs; the Web site of the Consortium of Social Science Associations [http://www.cossa.org] posts up-to-date information on the scientific community's response to this matter).

Misrepresentation of Research Findings

A recent report of the Union of Concerned Scientists (UCS, 2004) documents attempts at the federal level to distort and suppress scientific findings that run counter to a conservative social agenda. These efforts have encompassed a broad range of areas, including air pollution, heat-trapping emissions, drug-resistant bacteria, environmental issues, military intelligence, and reproductive health.

But efforts at the federal level are only the tip of the policy iceberg. There is a complex web of corporate and religious advocacy groups that operate at federal, state, and local levels to misrepresent scientific findings to policy makers and the public. Here we focus on findings related to reproductive health in general, and efforts to portray abortion as a threat to women's health in particular.

Constructing Abortion as a Threat to Women's Health

Concern for women's health has emerged as a central thrust for political forces attempting to overturn *Roe v. Wade* (1973) with the claim that the court failed to balance its concern for the negative effects of unwanted pregnancy with the alleged "fact" that abortion is detrimental to women's health—physical and mental. This campaign involves a sophisticated manipulation and misrepresentation of scientific findings to argue for legislation mandating informed consent scripts. Abortion is claimed to cause infertility, breast cancer, and mental disorder, among other ills. In particular, abortion is described as causing severe and long lasting posttraumatic stress (labeled "postabortion syndrome" or PAS). Meanwhile, the relative risks of having an unwanted birth are minimized (American Medical Association [AMA], Council on Scientific Affairs, 1992; David, Dytrych, & Matejeck, 2003; Russo, 1992). The short-term goal of this organized campaign is to deter women from having an abortion. The ultimate goal is overturning *Roe v. Wade*.

But even if *Roe v. Wade* (1973) stands, postabortion syndrome is designed to be used as a rationale for suing physicians for psychological damage alleged to result from having an abortion. A goal is to deter physicians from providing abortions, and the strategy involves attempts to enact legislation making physicians criminally liable for such "damages" (hence, not covered by malpractice insurance; Lee, 2003; Reardon, n.d.; and Russo & Denious, 1998 document these efforts).

There is no scientific basis for constructing abortion as a severe physical or mental health threat (e.g., Adler et al., 1990, 1992; AMA Council On Scientific Affairs, 1992; Denious & Russo, 2000; National Academy of Sciences, 1975; Russo, 1992; Schwartz, 1986). The power of expectancy effects is well documented, however (Rosenthal, 2002). In addition to subverting the informed consent process, misrepresenting the dangers of abortion may compound the stress of unwanted pregnancy by damaging a woman's belief in her ability to cope with her decisions, distorting her appraisals of events, and emotionally charging her experience. As a result, abortion can serve as a lightning rod for negative emotions originating in preexisting or concurrent conditions. Consequently, women may be led away from dealing with the deeper issues stemming from such conditions (see

Rubin & Russo, 2004, for a more extended discussion of these issues and their implications for the integrity of the therapeutic process).

The importance of providing patients with informed consent about any medical procedure is inarguable, but—as emphasized in the amicus briefs of the American Psychological Association on these issues—the usefulness of mandated scripts that contain erroneous information and that may or may not apply to a particular woman's circumstances or be written in terms that she can understand is highly questionable (see *Akron v. Akron Center for Reproductive Health, Inc.*, 1983; *Bowen v. Kendrick*, 1988; *Planned Parenthood of S.E. Pennsylvania v. Casey*, 1992; *Thornburgh v. American College of Obstetricians and Gynecologists*, 1986). More significantly, corrupting the informed consent process through mandating scripts that do not accurately represent the best of the scientific literature sets an alarming precedent. Although these strategies have been used primarily to manipulate the informed consent process related to abortion, legislating mandated scripts to dictate the interaction between health provider and patient has broader implications. Such activities are of concern to all, but particularly to members of ethnic groups who have historical reasons for mistrusting the intentions of medical researchers and practitioners (Bird & Bogart, this issue).

Research findings can and should be applied rationally to inform public policies in beneficial ways. However, the words of Illinois State Representative Dan Reitz illustrate the importance of establishing a norm of respect for research findings if they are to be used effectively. When Reitz was criticized for introducing a mandatory informed consent bill with misleading content he responded: "I'm not really sure about the science . . . . My intent was strictly about limiting abortion" ("Abortion: Making scare tactics legal," 2001, p. 6).

Many women may experience the pressures of society's "motherhood mandate," i.e., the expectation that all women should be mothers and devote themselves to raising their children (Russo, 1976), and may feel stigmatized for being or choosing to remain childless (see Lampman & Dowling-Guyer, 1995; Park, 2002). In the process, a woman likely considers the relationship of her choices to the values and norms enforced by others, even if only to resist them. As Stanworth (1987) has observed, it is misleading to evaluate reproductive technologies in the abstract, and it is similarly difficult to "find a position on motherhood from which we can say clearly and unambiguously what women want or need" (p. 3–4). Whether a woman is seeking an abortion or deciding to pursue costly, invasive infertility treatments, she must assess the costs and benefits of her options and deserves the best information available to inform this personal decision. But instead, she is told that "women who choose abortion suffer significantly more physical and psychological health problems than those who give birth" and that abortion is a "poor choice" for every woman (Elliot Institute, 2002, p. 1; available at http://www.afterabortion.org/news/poorchoice_release.html; see www.PoorChoice.org for information about the profile "poor choice" campaign).

Initially, the absence of scientific evidence for widespread and severe negative mental health effects of abortion hampered efforts of pro-life advocates to influence legislators and policy makers, many of whom use science to inform their positions and decisions. The strategy has thus shifted from argument by anecdote and clinical example to systematic misrepresentation of findings and attempts to undermine the credibility of researchers who refute their claims. Most worrisome is the publication of deeply flawed studies that contain miscoded data and meaningless findings (e.g., Reardon & Cougle, 2002) which are then used as "evidence" that abortion is harmful to women.

Dissemination of Results Needs Monitoring

Ethical use of research to inform public policy requires accurate presentation of findings. But even if findings are appropriately described in a scientific journal, this does not mean that they will be accurately disseminated. A recent example of how research findings are being distorted in the process of dissemination is found in a press release by the Elliot Institute (2004) with the headline "Death Rate Of Abortion Three Times Higher Than Childbirth . . . ." (Circulated by einews@afterabortion.info March 5, 2004). The release, which described findings of a Finnish records linkage study appearing in the American Journal of Obstetrics and Gynecology (Gissler et al., 2004), reported the finding that the mortality rate of abortion was 2.96 times that for pregnancies carried to term, giving the impression that having an abortion is more dangerous than giving birth. But in the article itself, those researchers had examined causes of deaths from different sources and reported that after excluding deaths for medical reasons unrelated to the abortion, the pregnancy-associated death rate for abortion was lower than that for giving birth.

Similarly, our own work (Russo & Denious, 2001) reported a significant correlation between abortion and depression, but observed that the relationship was no longer significant when relevant variables, including histories of childhood abuse and partner violence, were controlled. To our surprise, our findings have been misrepresented by pro-life advocacy groups in support of the argument that having an abortion subsequently puts women at greater risk of intimate partner violence due to abortion-induced feelings of anger and trauma on the part of both the woman and her partner (see Reardon et al., 2002).

The dissemination of scientific findings on government Web sites has also been politically manipulated. For example, the Bush administration had the National Cancer Institute (NCI) remove its fact sheet, which reported a lack of relationship between abortion and breast cancer and explained that after previous uncertainty, the findings of several well-designed studies published in the mid-1990's had resolved the issues (NCI, 2002a). The NCI's Web site was then revised to suggest that the issue of a link between abortion and breast cancer had studies of equal weight on both sides (NCI, 2002b).

Subsequently, under pressure from Congress, the NCI convened a panel to review the issue, with the NCI board of scientific counselors ultimately unanimously concluding that the evidence for the absence of an abortion/cancer link was "well-established" (NCI's highest standard) (Boonstra, 2003a, p. 3). The agency once again updated its Web site affirming the lack of relationship (NCI, 2003c). Nonetheless, efforts to link abortion to breast cancer continue in the minds of women.

Politicization of issues related to sexuality and reproduction that has led to mixing science and politics is reflected, also, in the misrepresentation of the effectiveness of "old fashioned" birth control technology (i.e., the male condom). As part of a campaign to promote abstinence outside marriage, scientific evidence has been distorted to undermine public confidence in condoms. Despite the compelling body of evidence that condoms protect against HIV and STIs, political forces continue to assert condoms are ineffective in preventing STIs (Cates, 2001). In the process, both the Centers for Disease Control (CDC) and the State Department's Agency for International Development (USAID) have had to revise their Web sites to emphasize a lack of evidence for condom effectiveness (Boonstra, 2003a).

Several lessons can be drawn from these activities. First, conscientiously assembling and evaluating a large body of scientific evidence is not enough. Scientists must also be involved in dissemination of their research findings beyond simple publication in scientific journals and actively work to ensure that the findings are not misrepresented to the public. Given the need for rapid response to misrepresentations of science in the popular media, the development of credible and thoughtful Web-based outlets for scientific comment is needed, such as those sponsored by the Society for the Psychological Study of Social Issues (http://www.asap-spssi.org), the Society for the Psychology of Women (http://www.prochoiceforum.org), and the Alan Guttmacher Institute (http://www.guttmacher.com).

Second, thoughtful analysis and complex explanation do not make for compelling sound bites or eye-grabbing headlines. Accurate presentation of the conclusions becomes overshadowed by conflicting interpretations of findings and the creation of "headline truths." As various reproductive technologies become more widespread (but remain controversial), it will be important to establish opportunities and outlets to counter myths and misinformation spread by inaccurate reporting. Thus, the public information offices of scientific associations such as the American Psychological Association or the American Association for the Advancement of Science, and evidence-based policy organizations such as the Alan Guttmacher Institute have key roles to play in ensuring that policymakers and the public receive accurate renditions of the state of scientific research in any particular area.

Third, the anti-sexuality values and attitudes that underlie the campaign against condoms also underlie resistance to other forms of birth control, such as abortion and emergency contraception. The view is that if birth control eliminates the threat of unwanted pregnancy, that threat can no longer serve as a deterrent to sexual behavior outside of marriage. Thus, educating the public about the fact that emergency contraception (EC) does not involve abortion will not eliminate resistance to EC adoption, unless research findings can be disseminated that counter the assertion that availability encourages irresponsible sexual behavior (Sherman, this issue).

Conclusion

Reproductive technologies—old and new—raise a host of social and legal issues that challenge some people's values and create intense controversy. In the current politicized context it is important for public policies to be informed by theory-based research and advocated by individuals and groups committed to a rational and sensitive discussion of the issues. Thus, psychologists in their roles as researchers, practitioners, and public interest advocates must be active in shaping and implementing policies in this arena.

In particular, research that is methodologically sophisticated and directed at resolving specific policy questions should be instrumental in informing the policy-making process related to specific technologies. The resulting knowledge base can be used to debunk myths about a technology (e.g., the idea that women use abortion to avoid having children when most adult abortion patients are already mothers; Russo, Horn, & Schwartz, 1992). Psychological research can help provide more accurate informed consent, particularly with regard to behavioral, psychological, and social outcomes. Knowledge about the range of outcomes and the factors that contribute to their development can help establish clear and realistic expectations, as well as reduce the stress associated with use of a particular technology. Despite the fact that research findings can inform the policy process, efforts to suppress and distort scientific findings in the service of a particular social agenda have undermined the integrity of the scientific advisory process and can damage the public's understanding of important reproductive health issues. In this context, in addition to conducting high quality policy-relevant research, researchers have an ethical responsibility to ensure that their research findings are disseminated and interpreted appropriately. In the final analysis, reproductive technologies as a means to empower women and help them achieve their reproductive goals have yet to fulfill their promise.

References

of Abortion: Making scare tactics legal. (2001, March 12). *Newsweek, CXXXVII*, 11, 6.

Adler, N. E., David, H. P., Major, B. N., Roth, S. H., Russo, N. F., & Wyatt, G. E. (1990). Psychological responses after abortion. *Science, 248*, 41–44.

Adler, N. E., David, H. P., Major, B. N., Roth, S. H., Russo, N. F., & Wyatt, G. E. (1992). Psychological factors in abortion: A review. *American Psychologist, 47*, 1194–1204.

Akron v. Akron Center for Reproductive Health, Inc., 462 U.S. 416 (1983).

American Medical Association, Council on Scientific Affairs. (1992). Induced termination of pregnancy before and after *Roe v. Wade*: Trends in the mortality and morbidity of women. *Journal of the American Medical Association, 268*, 3231–3239.

Boonstra, H. (2003a). Critics charge Bush mix of science and politics is unprecedented and dangerous. *The Guttmacher Report on Public Policy, 6*(2), 1–2.

Boonstra, H. (2003b). U.S. AIDS policy: Priority on treatment, conservatives' approach to prevention. *The Guttmacher Report on Public Policy, 6*(3), 1–3.

Bowen v. Kendrick, 487 U.S. 589 (1988).

Cates, W., Jr. (2001). The NIH condom report: The glass is 90% full. *Family Planning Perspectives, 33*, 231–233.

Cohen, S. (2002). Elections make drive for reproductive health and rights an even steeper uphill battle. *The Guttmacher Report on Public Policy, 5*(5), 1–2.

David, H. P., Dytrych, Z., & Matejcek, Z. (2003). Born unwanted: Observations from the Prague Study. *American Psychologist, 58*, 224–229.

Denious, J. E., & Russo, N. F. (2000). The socio-political context of abortion and its relationship to women's mental health. In J. Ussher (Ed.), *Women's Health: Contemporary International Perspectives* (pp. 431–439). London: British Psychological Society.

Drazen, J. M., & Ingelfinger, J. R. (2003). Grants, politics, and the NIH. *New England Journal of Medicine*, 2259–2261.

Elliot Institute. (2003). "New Poor-Choice Rhetoric Exposes Abortion's Dangers." Press release retrieved on December 10, 2004 from http://www.afterabortion.org/news/poorchoice_release.html

Elliot Institute. (2004, March 5). Death rate of abortion three times higher than childbirth: 13-year population study published in top OB/gyn journal. Press release retrieved December 10, 2004 from http://www.afterabortion.org/news/GisslerAJOG.htm

Firestone, S. (1970). *The dialectic of sex*. New York: William Morrow and Company, Inc.

Gissler, M., Berg, C., Bouvier-Colle, M., & Buekens, P. (2004). Pregnancy-Associated mortality after birth, spontaneous abortion, or induced abortion in Finland, 1987–2000. *American Journal of Obstetrics and Gynecology, 190*, 422–427.

Gordon, L. (1990). *Woman's body, woman's right: Birth control in America*. New York: Penguin Books. (Original list published 1976).

Holloway, M. (2001, April). Aborted thinking: Reenacting the Global Gag Rule threatens public health. *Scientific American, 284*(4), 19–20.

Lampman, C., & Dowling-Guyer, S. (1995). Attitudes toward voluntary and involuntary childlessness. *Basic and Applied Social Psychology, 17*, 213–222.

Lee, E. (2003). *Abortion, motherhood, and mental health*. New York: Aldine de Gruyter.

Leshner, A. I. (2003). Don't let ideology trump science. *Science, 302*(28), 1479.

Lindblom, C. E., & Woodhouse, E. J. (1993). *The policy making process* (3rd ed.). Englewood Cliffs, NJ: Prentice Hall.

National Academy of Sciences. (1975). *Legalized abortion and the public health*. Washington, DC: National Academy Press.

National Cancer Institute. (2002a). *Abortion and breast cancer*. Retrieved December 10, 2004, from http://www.democrats.reform.house.gov/Documents/20040817143732_39165.pdf

National Cancer Institute. (2002b). *Early reproductive events and breast cancer*. Retrieved December 10, 2004, from http://www.democrats.reform.house.gov/Documents/30040817/4380743596.pdf

National Cancer Institute. (2003c). *Abortion, miscarriage, and breast cancer risk*. Retrieved December 10, 2004, from http://cis.nci.nih.gov/fact/3_75.htm

Park, K. (2002). Stigma management among the voluntarily childless. *Sociological Perspectives, 45*, 21–45.

Planned Parenthood of S.E. Pennsylvania v. Casey, 505 U.S. 833 (1992).

Reardon, D. C. (n.d.). *The Jericho Plan: Breaking down the walls which prevent post-abortion healing*. Springfield, IL.: Acorn Books.

Reardon, D. C., & Cougle, J. R. (2002). Depression and unintended pregnancy in the National Longitudinal Survey of Youth: A cohort study. *British Medical Journal*, 151–152.

Reardon, D. C., Ney, P. G., Scheuren, F., Cougle, J., Coleman, P. K., & Strahan, T. W. (2002). Deaths associated with pregnancy outcome: A record linkage study of low income women. *Southern Medical Journal, 95,* 834–841.

Reppert, B. (2004, January 06). Politics in the lab hits US scientific integrity. *Christian Science Monitor.* Retrieved May 29, 2004, from http://www.csmonitor.com/2004/0106/p11s02-coop.html

Roe v. Wade, 410 U. S. 113 (1973).

Rosenthal, R. (2002). Covert communication in classrooms, clinics, courtrooms, and cubicles. *American Psychologist, 57,* 839–849.

Rubin, L., & Russo, N. F. (2004). Abortion and mental health: What therapists need to know. *Women & Therapy, 27*(3/4), 69–90.

Russo, N. F. (1976). The motherhood mandate. *Journal of Social Issues, 32,* 143–154.

Russo, N. F. (1992). Psychological aspects of unwanted pregnancy and its resolution. In J. D. Butler & D. F. Walbert (Eds.), *Abortion, medicine, and the law* (4th ed., pp. 593–626). New York: Facts on File.

Russo, N. F., & Denious, J. E. (1998). Why is abortion such a controversial issue in the United States? In L. J. Beckman & S. M. Harvey (Eds.), *The new civil war: The psychology, culture, and politics of abortion* (pp. 25–60). Washington, DC: American Psychological Association.

Russo, N. F., & Denious, J. E. (2001). Violence in the lives of women having abortions: Implications for practice and public policy. *Professional Psychology: Research and Practice, 32,* 142–150.

Russo, N. F., Horn, J., & Schwartz, R. (1992). Abortion in context: Characteristics and motivations of women who seek abortion. *Journal of Social Issues, 48,* 182–201.

Schwartz, R. A. (1986). Abortion on request: The psychiatric implications. In J. D. Butler & D. F. Walbert (Eds.), *Abortion, medicine, and the law* (3rd ed., pp. 323–340). New York: Facts on File.

Solinger, R. (Ed.). (1998). *Abortion wars: A half century of struggle, 1950–2000.* Berkeley, CA: University of California Press.

Stanworth, M. (Ed.). (1987). *Reproductive technologies: Gender, motherhood, and medicine.* Minneapolis, MN: University of Minnesota Press.

Thornburgh v. American College of Obstetricians and Gynecologists, 476 U.S. 747 (1986).

Union of Concerned Scientists. (2004, March 26). *Scientific integrity in policymaking.* Retrieved , from http://www.ucsusa.org/global_environment/rsi/page.ctm/pageID=1449

Weiss, R. (2004, February 28). UCSF scientist dropped from bioethics council. *San Francisco Chronicle,* A1, A10.

NANCY FELIPE RUSSO is Regents Professor of Psychology and Women's Studies at Arizona State University. A former member of the Task Force on Postabortion Emotional Responses of the American Psychological Association, she is the current co-chair of the Task Force on Reproductive Issues of the Society for the Psychology of Women.

JEAN E. DENIOUS received her doctorate in social psychology from Arizona State University in 2004, and currently teaches in the Department of Psychology and in the Women's Studies program at ASU. Her research and teaching interests are in gender and health, with emphases on body image and related regulatory processes, and in social, psychological, and political contexts of reproductive health issues.

Journal of Social Issues, Vol. 61, No. 1, 2005, pp. 193–205

Generation of Knowledge for Reproductive Health Technologies: Constraints on Social and Behavioral Research

Cynthia Woodsong*

Family Health International

Lawrence J. Severy

University of Florida and Family Health International

Advances in new reproductive health technologies have surfaced an array of social and behavioral issues regarding decision-making and use of these technologies, underscoring the need for research on such topics as reproductive health decision-making, sexual practices, and norms and values for childbearing and family formation. Using topical microbicides as an example of a new method to prevent Sexually Transmitted infections (STIs), Human Immunodeficiency Virus (HIV) and/or pregnancy, we focus on gaps in information to inform reproductive health decision-making, noting in particular the discrepancies between data on clinical efficacy and typical use-effectiveness. Constraints on government and private sector support for research, particularly research on aspects of sexual behavior, contribute to problems with the availability of information for decision-making about use of reproductive health technologies.

An array of reproductive health technologies has grown steadily over the past few decades, with an accompanying array of unresolved issues regarding decision making and use of these technologies. In particular, both users and practitioners of health services need reliable and accessible information about the increasing number of reproductive health options. It is not enough for couples to be advised about the clinically-proven safety and efficacy of reproductive health technologies.

*Correspondence concerning this article should be addressed to Cynthia Woodsong, P.O. Box 13950, Research Triangle Park, North Carolina, 27709 [e-mail: Cwoodsong@fhi.org].

The authors wish to thank reviewers for their insights and advice. The opinions expressed in this article do not necessarily reflect the policies of Family Health International or the University of Florida.

They benefit, also, from frank and open discussions of how methods fare under typical use, and these discussions must be informed by findings from social and behavioral research on such topics as reproductive health decision making, sexual practices, and norms and values for childbearing and family formation. However, support for public research and privately funded research on such topics is currently quite tenuous.

In this article, we discuss the need for and problems with the availability of reliable research-based information to help individuals and couples, as well as their health practitioners, make decisions about use of reproductive health technologies. Using topical microbicides as an example of a new method to prevent STIs, HIV, and pregnancy, we focus on gaps in information, noting in particular the discrepancies between data on clinical efficacy and typical use-effectiveness. We next consider government and then private sector support for the generation of social and behavioral science knowledge about reproductive health. We conclude by arguing that constraints on research, particularly research on aspects of sexual behavior, negatively impact the availability of information for decision-making about use of reproductive health technologies.

Information Needed for Reproductive and Sexual Health

Reproductive health technologies include drugs, devices, and medical interventions that control reproduction and/or prevent sexually transmitted infections (STIs), such as contraceptives and products used to enhance fertility, as well as techniques for in vitro and in vivo fertilization. Other reproductive health products are intended to help people avoid some of the undesired outcomes of sexual activity, such as emergency contraceptives to prevent an unwanted pregnancy and topical microbicides to prevent STIs, including HIV (Beckman & Harvey, this issue). And finally, as populations age, technologies are increasingly needed to manage the physiological and emotional transitions associated with menopause (Naughton, Jones, & Shumaker, this issue).

Issues associated with access to reproductive health resources have a long history. In particular, international conferences in Cairo (1994) and Beijing (1995) built strong international support for the proposition that couples and individuals should have the information and means to help them achieve their desired number of children (Gillespie, 2004; United Nations, 2001). More recently, the World Health Organization (WHO, 2002) moved to endorse the principle of "sexual health." Like the conceptualization of health as more than absence of disease, sexual health is viewed as "a state of physical, emotional, mental and social well-being related to sexuality" (WHO, 2002, p. 2). The concept of sexual health further affirms the human rights of all individuals to access sexual and reproductive health care services and information related to sexuality, including information that can aid in deciding if, when, and how to have children.

Good data, tools, and information are needed to help people understand the risks and benefits associated with different reproductive health choices. Certainly, people need a thorough description of assisted reproductive health technology options, their protocols for use, side effects, and efficacy. However, in the process of decision making, potential users will consider this information and interpret it through their own values and belief systems, taking into consideration both medical and socio-behavioral aspects (Woodsong, Shedlin, & Koo, 2004).

In addition to basic information needed for decision making regarding use of reproductive health technologies, data on sub-groups is important. Bachrach and Abeles (2004) have observed that important basic social science concepts, such as socio economic and demographic indicators are often used "superficially and mechanically" (p. 23). Harvey and Nichols (this issue) observe that women with higher levels of education are more likely to choose "medical abortion" using pharmaceutical drugs over surgical methods such as vacuum aspiration or dilation and curettage. The observation that women with more limited economic means choose the more costly and more invasive surgical options raises questions that require an examination of the larger contextual environment that influences their decisions. Inferring simple causality of economic status on method choice is clearly inadequate.

And, for some assisted reproductive technologies, legal considerations associated with medical options create an additional layer of complexity for decision making. Ciccarelli and Ciccarelli (this issue) point out that laws which govern parental rights and obligations vary between states and this has important implications for those considering assisted reproductive technologies for having children. They believe that such medical advances "have greatly outpaced society's, and consequently the law's, ability to address the relationships and attendant rights and responsibilities which arise between the parties" involved (pg. 127). Information on family formation and multi-cultural family values is needed to effectively advise people on their reproductive health decisions for assisted fertility, as well as shape the development of related health policies.

This is complicated by the climate of cultural diversity in the United States. The current percentage of adults in the United States who are members of a racial minority group, is expected to increase from 23% in 2000 to 40% in 2025 (U.S. Census Bureau, 2001). As these numbers grow, so will the need for an understanding of varying norms for reproductive health, including family formation and sexual practices among minorities. Individuals and couples from minority groups in the United States may conceptualize family and parenthood differently than those in mainstream society.

Furthermore, both mainstream and minority experiences of "family" may be out of synch with the politico-legal system. For example, political concern about family and household structure has, over the past few decades, included legislation and programs designed to defend and protect the institutions of heterosexual marriage and dual-parent households. Since only slightly more than half (53%) of U.S. households meet this standard (U.S. Census Bureau, 2002), a significant number of households may not be effectively embraced by services that cater to this ideal. This includes the increasing numbers of single parent and same-sex partner households who may have reproductive health goals for childbearing.

People's ability to limit their fertility is also subject to shifts in the political environment (Hwang & Stewart, 2004; Russo & Denious, this issue). Sherman (this issue) observes that in the United States, "one half of all pregnancies are unintended, and half of all unintended pregnancies end in abortion..." (p. 141). Yet, anti-abortion advocacy continues to dominate the political landscape. One consequence has been strident debates over emergency contraceptives, with powerful advocacy groups labeling these as abortifacients (Gervais & Miles, 1990; Klitsch, 1990). Although concerned public health researchers have clearly demonstrated that emergency contraceptive use does not necessarily disrupt an implanted embryo (Gervais & Miles, 1990; Piaggio, von Hartzen, Grimes, & Van Look, 1999), sensationalized reporting has cast a long-politicized shadow on emergency contraceptives, an important addition to the reproductive health pharmacopeia.

In summary, the great variability in values for childbearing and family formation and sexual practices and meanings is not well accommodated by health services and regulations to limit or extend childbearing. Research is needed on values and norms for sexual health and family formation. Such research could inform policies for, and delivery of, reproductive health technologies. However, as others in this issue have noted (Ciccarelli & Beckman, this issue; Ciccarelli & Ciccarelli, this issue), the much-needed information that would aid both providers and consumers of assisted reproductive technologies is not available.

Effectiveness Equipoise

One area where the lack of reliable reproductive health data is particularly problematic concerns methods to prevent pregnancy and/or STIs, including HIV. A basic concept in medical research ethics is equipoise, which describes a situation where there is reasonable doubt in the scientific community about the relative benefits of different medical interventions, such that one is not considered to have a clear advantage over the other (Freedman, 1987; London, 2001). Since health practitioners are obligated to recommend the "best" method or treatment approach available, there must be equipoise in order to ethically justify clinical trials designed to test different interventions.

We consider that a practitioner's judgment of what is best should be informed by an understanding of not only clinical efficacy, but also of typical use-effectiveness. For example, it does little good to recommend a method for prevention of pregnancy or STI transmission that is 98% effective, if only 20% of typical users can or will use it correctly and consistently (Severy & Newcomer, this issue; Spieler, 1997). Even though the controlled environment of a clinical trial may demonstrate that a product used consistently and correctly has a certain rate of efficacy, in real-life circumstances, use-effectiveness may be much lower (Elias & Coggins, 2001; Hearst & Chen, 2004).

Information on effectiveness under typical use, which would include variations in sexual behavior and practices, could help inform health professionals' advice. "User-failure" and "typical use" rates are available in the health literature (Fu, Darroch, Haas, & Ranjit, 1999; Trussell & Vaughn, 1999) but reliable information on the sexual practices and circumstances that contribute to such failure is scant (Brown & Eisenberg, 1995; Celentano, 2004; Miller, 1986). Furthermore, the limited reliable information on typical use can be difficult to apply to non-typical users, including those with values that differ from mainstream society. Given the range of variability in actual use-effectiveness, health practitioners may be limited in their ability to judge which prevention method to recommend to clients who request advice about methods to prevent pregnancy and/or STIs. We consider, then, that a situation of effectiveness equipoise exists for many products for prevention of pregnancy and/or STIs.

Topical Microbicides

Gaps in health practitioners' knowledge about sexual behavior may contribute to problems with the uptake of new reproductive health technologies. This situation could easily occur with topical microbicides, when they become available. Topical microbicides are chemical substances formulated for insertion into the vagina or rectum to prevent or reduce transmission of sexually transmitted infections, including HIV, as well as pregnancy (but not all microbicides will also be contraceptive; Harrison, Rosenberg, & Bowcut, 2003). Different mechanisms of action are currently being developed and tested (Moore & Shattock, 2004; Koo, Woodsong, Dalberth, Viswanathan, & Simmons-Rudolph, this issue), with over 60 different products currently in the development pipeline (Harrison et al., 2003).

Efficacy and effectiveness questions have already surfaced as a potential problem for the introduction and use of topical microbicides. For example, the first generation of microbicides may demonstrate only a 30% efficacy against HIV infection (Harrison et al., 2003; Rockefeller Foundation, 2002a; Rustomjee & Abdool Karim, 2001). This will not compare favorably to condom efficacy and thus a practitioner, assuming that equipoise is not obtained, may be reluctant to recommend them over condoms (Severy & Newcomer, this issue).

Similarly, microbicides' expected lower efficacy concerns those who support and fund condom programs and who argue that substituting microbicides for condoms will result in increased infections (Rockefeller Foundation, 2002b). However, since problems with the validity and reliability of self-reported condom use and sexual behavior are well documented (Catania, Gibson, Chitwood, & Coates, 1990; Weir, Roddy, Zekeng, & Ryan, 1995; Zenilman et al., 1999), we argue that effectiveness equipoise will likely be present. In this situation, health practitioners would benefit from findings of social and behavioral research on the acceptability of microbicides compared with condoms. This could provide important information to help assess a client's ability to correctly and consistently use condoms.

The concept of "partial effectiveness" (Halloran, Longini, Jr., Haber, Struchiner, & Brunet, 1994; Severy & Newcomer, this issue) presents a further complication that illustrates the potential for effectiveness equipoise for microbicides. If a microbicide becomes available that is only 30% to 50% effective, how should public health practitioners advise their clients? Since condoms are also a partially effective method, how can practitioners convey in a meaningful way what is meant by partial effectiveness of microbicides compared to condoms, if true use-effectiveness of both methods is not known? Errors of judgment in conveying this information could result in users' mistaken interpretations that, in turn, could lead to disinhibition of risk reduction behaviors and increased infections.

As stated above, the first generation of microbicide products will likely be less effective than subsequent products. If initial microbicide products become associated with infections, efforts to test and produce more effective products could be hampered by lack of confidence in the product concept. As Gita Ramjee commented in her summary of social science issues raised at the Microbicides 2004 meetings in London (Ramjee, 2004) "the field can not tolerate too many efficacy failures." People's perceptions of both safety and efficacy will influence acceptability and use, which in turn will influence the demand for microbicides (Heise, 1997).

Issues in microbicide development highlight the need for information to help guide providers and consumers in making choices to achieve good sexual and reproductive health. We have noted that lack of research-generated knowledge could influence supply, demand, and correct use of reproductive health options. We now briefly consider issues that exacerbate problems with the generation of knowledge based on social and behavioral research, including influences of the political climate. Research conducted with both public and private sector support is affected by these issues, albeit in different ways.

The Research Enterprise

Constraints in Public Sector Funding

According to the National Bioethics Advisory Commission (NBAC), "in the past two decades, phenomenal growth has occurred in federally and industry-sponsored biomedical research. Federal expenditures for medical and health research conducted in the United States and in foreign countries almost doubled from $6.9 billion to $13.4 billion between 1986 and 1995" (NBAC, 2001, p. 4). However, social and behavioral research in both the public and private sector receives much less support than clinical and biomedical research. For example, although the National Institutes of Health (NIH) funding for social and behavioral research and training has been increasing, it accounts for only 10% of the total budget (Bachrach & Abeles, 2004).

To some extent, lower levels of funding are due to the comparatively high costs of conducting biomedical studies. However, it can be argued that inherent biases in dominant paradigms for biomedical research may overshadow social and behavioral research (Dean, 2004; Jones, 2004). In short, clinicians may not value behavioral research as much as what they consider to be "hard science." It is not our purpose in this article to describe and defend the basic premises of research paradigms. We do, however, consider that one consequence of the dominance of biomedical science is less financial support for social and behavioral research, and insufficient recognition of the contributions that such research can make to improving health.

Another factor contributing to limited flows of funding for social and be-havioral research is that some topics are currently considered as controversial. Recently, U.S. researchers have been advised (informally) to carefully scrutinize their requests for government research funding support and avoid a number of words commonly used to refer to sexual behaviors since these could negatively influence funding decisions (Goode, 2003; Kaiser, 2003; Stewart, 2003). In this environment, some have observed (Blackburn, 2004, Nature Immunology, 2003) that scientists may become less willing to serve at NIH. *The Lancet* ("Keeping Sci-entific," 2002) notes that positions on scientific advisory panels are increasingly being filled with individuals who support partisan, rather than scientific, views on reproductive health issues. One consequence could be less social and behavioral research, since it could become much safer for researchers to propose work on less controversial topics, and safer for funding decision-makers to approve them.

In addition to efforts to generate new knowledge, reproductive health research that has been conducted, published, and accepted by the scientific community is targeted by political groups for censure (Drazen & Ingelfinger, 2003; Russo & Denious, this issue). Information on such topics as condom safety and efficacy and abortion-related health issues has been misrepresented or removed from public access (Cates, 2002; "Keeping Scientific," 2002; Waxman, 2003;). This can be confusing to providers and consumers of health care information who may be unsure of the quality of reproductive health information, if they can even find it (Taubes, 1995).

Recent disclosure of U.S. congressional staff requests to identify scientists researching HIV/AIDS, human sexuality, and risk-taking behaviors created con-cern among the research community (Gallagher, 2003). Particularly alarming was news that the NIH contacted researchers for additional information that might be needed to defend funding support in an environment of increasing pressure to not fund studies of sexual behaviors (Drazen & Ingelfinger, 2003; Kaiser, 2003). This prompted the *New England Journal of Medicine* to publish an article in which the NIH scientific review process was carefully explained and defended (Drazen & Ingelfinger, 2003), a process that was also strongly endorsed and defended by the Association of American Medical Colleges. Nevertheless, the overall climate of uncertainty for federally funded research could create a disincentive to the conduct of research on sexual behaviors.

Constraints in Private Sector Research

In the private sector, a profit motive influences reproductive health research. Research and development for new reproductive health technologies is slow and extremely expensive and the potential economic benefits of such new technologies must be demonstrated in order to justify investments. Reproductive health products such as topical microbicides could benefit vast numbers of women in developing countries, but only if they are available at a very low cost (Moore & Shattock, 2004). Since it is not clear how or when the investment in research and development will be recouped, large pharmaceutical companies have not participated in development of microbicide products and, thus, product development is proceeding very slowly (Rockefeller Foundation, 2002a). Similarly, over the past few decades, the economic return on new contraceptive technologies has, for the most part, proven to be low or has incurred high litigious costs, contributing to a lack of industry interest (Mastroanni, Donaldson, & Kane, 1990).

The process of approving any new reproductive health technology for licensure in the United States is not only slow and, thus, costly, but it is susceptible to political influences that can further, if not indefinitely, delay a return on research investment (Gervais & Miles, 1990). For example, licensure of emergency contraceptives and deregulation for over-the-counter availability took years, in part due to earlier-mentioned controversies about the mechanisms of action with regard to pregnancy (Gervais, 1990; Bell & Mahowald, 2000; Klitsch, 1990). This created problems with marketing and distribution, as large retailers refused to carry the products (Canedy, 1999). This, in turn, likely cut into profit margins for the pharmaceutical companies. Such problems with development of new reproductive health technologies could result in less research to inform both product development and post-marketing research to inform the delivery of products that do make it through licensure. This is unfortunate since dissatisfaction with currently available contraceptive methods is associated with method discontinuation, which in turn contributes to high rates of unintended pregnancy (Fu et al., 1999).

Discussion

The application of research-based knowledge to policy and program service delivery is generally appreciated, although it is often difficult to achieve (Huberman, 1994; Lavis, Posada, Haines & Osei, 2002; Ulin, Robinson, Tolley, & McNeill, 2002). Program makers and policymakers need good data to work with, as well as advocates who bring the two together (Lomas, 2000). Researchers developing new reproductive health technologies frequently include only scant consideration of a limited range of social and behavioral factors. Although product developers may assume that "if you build it, they will come," issues with the uptake of new methods such as female condoms, Norplant, and contraceptive rings and patches demonstrate the importance of acceptability research (Ellertson & Winikoff, 1997; Severy & Newcomer, this issue). Many clinical researchers consider that behavioral acceptability research should wait until the post-marketing phase of product development, yet other scientists contend that behavioral research conducted concurrently with clinical trials can contribute to early acceptability and uptake as well as long-term correct and consistent use (Cottingham, 1997; Ellertson & Winikoff, 1997; Mauck, Rosenberg, & Van Damme, 2001;).

Historically, panels of scientific experts, drawing on the best of what research has to offer, are called upon to advise on the development of health programs and services. As mentioned earlier, the reluctance of many biomedical researchers to embrace social and behavioral studies is long standing and the comparatively low level of funding for such work can constitute an uphill climb for social and behavioral researchers. The current political environment further adds to the difficulties that such researchers face. Additionally, political, religious and philosophical considerations may affect the generation of such knowledge, as well as policymaker's and citizen's access to what does exist. We agree with Russo and Denious (this issue) that political ideologies, whether liberal or conservative, must not replace or delimit the generation of scientific knowledge designed to inform consumers and/or providers about reproductive health. All of the articles in this issue speak to our vulnerability to this threat. We argue that science should serve as both cornerstone and litmus test for decision making about reproductive health policy.

The range of current technologies available to protect reproductive health, and ensure or assist reproduction, is accompanied by a raft of social and political issues. Access to and choice of methods is shaped by factors outside the realm of formal medicine. These factors include not only the political and service delivery environment, but also individual and cultural values and beliefs about the available choices, including perceptions of what is "natural and right," a complex cosmology of religious, spiritual, and traditional cultural views (Woodsong et al., 2004). Such views, in effect, form a translative lens for health information and shape the decision-making process and thus the outcome—choice of method to achieve the given reproductive health goal.

Given the existence of this lens, it is imperative that accurate and comprehensible information be accessible to users of reproductive health care services. It must also be available to health care practitioners who advise users on their choice of services (Lavis, Posada, Haines, & Osei, 2004). Furthermore, in advising clients, practitioners should not rely simply on what is proven possible in clinical trial settings, but also consider behavioral and social factors influencing use-effectiveness. Access to good information to aid in the decision-making process can be problematic, in part because sources for funding potentially bias the type and nature of reproductive health research that is conducted. In the United States, there are currently uncertainties about the degree to which government-supported research on a variety of reproductive health topics will be supported through the life of a study, and how widely the knowledge generated will be used (Drazen & Ingelfinger, 2003).

In this article, we have considered some of the implications of political will on the generation of reproductive health knowledge and information. We agree with others that the science agenda for reproductive health must include not only technological advances, but also social and behavioral research on the intended users of these technologies (Bachrach & Abeles, 2003; Dean, 2004; Gillett, 2004). Policies that guide programs and services could be improved through application of research on user-perspectives to help remove effectiveness equipoise, so that couples and individuals can choose reproductive health technologies that best suit their needs and lifestyles.

While our focus here is on issues among U.S. populations, we do consider that a number of these issues are emergent in developing country settings. Although initial concerns about use of a new reproductive health method will likely center on ethical issues of equity and justice and the need to overcome issues of access to these technologies (WHO, 2001), we observe that mainstream cultural values for family formation and sexual practices and meanings will most certainly differ from those of the countries where assisted reproductive technologies are first developed and delivered. Recent studies of the demand for assisted reproductive technologies in developing country settings (Inhorn, 2003; Okonofua, 2003) demonstrate that as global demand increases, so will a need for attendant social and behavioral research, as described in this article.

References

Bachrach, C., & Abeles, R. (2004). Social science and health research: Growth at the National Institutes of Health. *American Journal of Public Health, 94*(1), 22–28.

Bell, B., & Mahowald, M. (2000). Emergency contraception and the ethics of discussing it prior to emergency. *Women's Health Issues, 10*(6), 312–316.

Blackburn, E. (2004). Bioethics and the political distortion of biomedical science. *New England Journal of Medicine, 350*(14), 1379–1380.

Brown, S., & Eisenberg, L. (Eds.). (1995). *The best intentions: Unintended pregnancy and the well-being of children and families.* Washington, DC: National Academy Press.

Canedy, D. (1999, May 14). Wal-Mart decides against selling a contraceptive. *The New York Times,* p. C1.

Catania, J., Gibson, D., Chitwood, D., & Coates, T. (1990). Methodological problems in AIDS behavioral research: Influences on measurement error and participation bias in studies of sexual behavior. *Psychological Bulletin, 108,* 339–362.

Cates, W. (2002). Let's not equivocate—Condoms work well. *Global AIDSLink, 73,* 14.

Celentano, D. (2004). It's all in the measurement: Consistent condom use is effective in preventing sexually transmitted infections. *Sexually Transmitted Diseases, 31*(3), 161–162.

Cottingham, J. (1997). Beyond acceptability: Users' perspectives on contraception: Introduction. In T. Ravindran, M. Berer, & J. Cottingham, (Eds.), *Beyond acceptability: User's perspectives on contraception* (pp. 1–4). London: Reproductive Health Matters/WHO.

Dean, K. (2004). The role of methods in maintaining orthodox beliefs in health research. *Social Science and Medicine, 58,* 675–685.

Drazen, J., & Ingelfinger, J. (2003). Grants, politics and the NIH. *New England Journal of Medicine, 349,* 2259–2261.

Elias, C., & Coggins, C. (2001). Acceptability research on female-controlled barrier methods to prevent heterosexual transmission of HIV: Where have we been? Where are we going? *Journal of Women's Health and Gender-based Medicine, 10,* 163–173.

Ellertson, C., & Winikoff, B. (1997). Why research on contraceptive user perspectives deserves public sector support: A free-market analysis. In T. Ravindran, M. Berer, & J. Cottingham (Eds.), *Beyond acceptability: User's perspectives on contraception* (pp. 15–22). London: Reproductive Health Matters/WHO.

Freedman, B. (1987). Equipoise and the ethics of clinical research. *New England Journal of Medicine, 317,* 141–145.

Fu, H., Darroch, J., Haas, T., & Ranjit, N. (1999). Contraceptive failure rates: New estimates from the 1995 national survey of family growth. *Family Planning Perspectives, 31*(2) 56–63.

Gallagher, R. (2003, December 1). No sex research please, we're Americans. *The Scientist,* 6.

Gervais, K., & Miles, S. (1990). *RU486: New issues in the American abortion debate.* Mineapolis, MN: Center for Biomedical Ethics.

Gillespie, D. (2004). Whatever happened to family planning and, for that matter, reproductive health? *International Family Planning Perspectives, 30*(1), 34–38.

Gillett, G. (2004). Clinical medicine and the quest for certainty. *Social Science and Medicine, 58,* 727–738.

Goode, E. (2003, April 18). Certain words can trip up AIDS grants, scientists say. *The New York Times,* p. A10.

Halloran, M., Longini I., Jr., Haber, M., Struchiner, C., & Brunet, R. (1994). Exposure efficacy and change in contact rates in evaluating prophylactic HIV vaccines in the field. *Statistics in Medicine, 13*(4), 357–377.

Harrison, P., Rosenberg, Z., & Bowcut, J. (2003). Topical microbicides for disease prevention: Status and challenges. *Clinical Infectious Diseases, 36,* 1290–1294.

Hearst, N., & Chen, S. (2004). Condom Promotion for AIDS Prevention in the developing world: Is it working? *Studies in Family Planning, 35*(1), 39–47.

Heise, L. (1997). Beyond acceptability: Reorienting research on contraceptive choice. In T. Ravindran, M. Berer, & J. Cottingham (Eds.), *Beyond acceptability: User's perspectives on contraception* (pp. 1–4). London: Reproductive Health Matters/WHO.

Huberman, M. (1994). Research utilization: The state of the art. *Knowledge and Policy: The International Journal of Knowledge Transfer and Utilization, 7,* 13–33.

Hwang, A., & Stewart, F. (2004). Family planning in the balance. *American Journal of Public Health, 94*(1), 15–18.

Inhorn, M. (2003). Global infertility and the globalness of new reproductive technologies: Illustrations from Egypt. *Social Science and Medicine, 56*(9), 1844.

Jones, K. (2004). Introduction to: Heresy and orthodoxy in medical theory and research. *Social Science and Medicine, 58,* 671–674.

Kaiser, J. (2003). Politics and biomedicine: Studies of gay men, prostitutes come under scrutiny. *Science, 300,* 403.

Keeping scientific advice non-partisan. (2002). *The Lancet, 360,* 1525.

Klitsch, M. (1990). *RU-486: The science and the politics.* New York: Alan Guttmacher Insitute.

Lavis, J., Posada, F., Haines, A., & Osei, E. (2004). Use of research to inform public policymaking. *Lancet, 364,* 1615–1621.

Lomas, J. (2000). Connecting research and policy. *Printemps, Spring,* 140–144.

London, A. (2001). Equipoise and International Human-Subjects Research. *Bioethics, 15,* 312–332.

Mastroanni, L., Donaldson, P., & Kane, T. (1990). *Developing new contraceptives: Obstacles and opportunities.* Washington, DC: National Academy of Sciences.

Mauck, C., Rosenberg, Z., & Van Damme, L. (2001). Recommendations for the clinical development of topical microbicides: An update. *AIDS, 15*(7), 857–868.

Miller, W. (1986). Why some women fail to use their contraceptive method: A psychological investigation. *Family Planning Perspectives, 18,* 27–32.

Moore, J., & Shattock, R. (2004). Preventing HIV-1 sexual transmission—Not sexy enough science, or no benefit to the bottom line? *Journal of Antimicrobial Chemotherapy, 52*(6): 890–892.

National Bioethics Advisory Commission. (2001). Ethical and policy issues in research involving human participants. *Ethical and policy issues in international research clinical trials in developing countries.* Bethesda, MD: Author.

Nature Immunology. (2003). The politics of federal research. *Nature Immunology 4*(12), 1151.

Okonofua, F. (2003). New reproductive technologies and infertility treatment in Africa. *African Journal of Reproductrive Health, 7*(1), 7–11.

Piaggio, G., von Hartzen, H., Grimes, D., & Van Look, P. (1999). Timing of emergency contraception with levonorgestrel of the Yuzpe regimen. *Lancet, 353*(9154), 721.

Ramjee, G. (2004, March 28–31). *Highlights and concluding remarks for Track C, Behavioural Science.* Presented at the Microbicides 2004 meetings, London.

Rockefeller Foundation. (2002a). *The science of microbicides: Accelerating development: A report by the Science Working Group of the Microbicide Initiative.* New York: Author.

Rockefeller Foundation. (2002b). *The public health benefits of microbicides in lower-income countries model projections: A report by the Public Health Working Group of the Microbicide Initiative.* New York: Author.

Rustomjee, R., & Abdool Karim, Q. (2001). Microbicide research and development—where to? *HIV Clinical Trials, 2*(3), 185–192.

Spieler, J. (1997). *Behavioral Issues in Dual Protection.* Paper presented at the Psychosocial Workshop, Population Association of America, Washington, DC.

Stewart, F. (2003). The war on words: Sensible compromise or slow suicide? *Contraception, 68,* 157–158.

Taubes, G. (1995). Epidemiology faces its limits. *Science, 26,* 164–169.

Trussell, J., & Vaughan, B. (1999). Contraceptive failure, method-related discontinuation and resumption of use: Results from the 1995 national survey of family growth. *Family Planning Perspectives, 31*(2), 64–93.

Ulin, P., Robinson, E., Tolley, E., & McNeill, E. (2002). Disseminating Qualitative Research. In P. Ulin, E. Robinson, E. Tolley, & E. McNeill, E. (Eds.), *Qualitative methods: A field guide for applied research in sexual and reproductive health* (pp. 187–191). Research Triangle Park, NC: Family Health International.

United Nations. (2001, 25–27 June). *United Nations Recommendations of the Expert Group Meeting. Application of Human Rights to Reproductive Sexual Health.* Geneva, Switzerland.

U. S. Census Bureau. (2001). *Census 2000 redistricting [Public Law 94-171] summary file.* Washington, DC: Author.

U.S. Census Bureau. (2002). Current Population Reports. *Statistical Abstract of the United States, 2002.* Washington, DC: US Government Printing Office.

Waxman, H. (2003, August). Politics and science in the Bush administration. US House of Representatives Committee on Government Reform—Minority Staff Special Investigations Division, Washington, DC.

Weir, S., Roddy, R., Zekeng, L., & Ryan, K. (1999). Association between condom use and HIV infection: A randomised study of self-reported condom use measures. *Journal of Epidemiological Community Health, 53*, 417–442.

World Health Organization. (2001). *Medical, ethical and social aspects of assisted reproduction.* Geneva, Switzerland: Author. Retrieved October 3, 2003, from http://www.who.int/reproductive-health/infertility/index.htm

World Health Organizaton. (2002). *Technical consultation on sexual health.* Geneva, Switzerland: Author. Retrieved April 17, 2004, from http://www.who.int/reproductive-health/gender/sexual_health.html

Woodsong, C., Shedlin, M., & Koo, H. (2004). Natural, normal and sacred: Beliefs influencing the acceptability of pregnancy and STI/HIV prevention methods. *Journal of Culture, Health and Sexuality, 6*(1), 67–88.

Zenilman, J., Weisman, C., Rompalo, A., Ellish, N., Upchurch, D., Hook, E., et al. (1999). Condom use to prevent incident STDs: The validity of self-reported condom use. *Sexually Transmitted Diseases, 22*, 15–21.

CYNTHIA WOODSONG earned a PhD in anthropology from the State University of New York in 1992 and then followed up with a post-doctoral fellowship in public health policy at the Carolina Population Center, University of North Carolina. Dr. Woodsong is a senior scientist in the Behavioral and Social Sciences Division at Family Health International (FHI) in Durham, North Carolina. At FHI, she is working on aspects of use-adherence and acceptability of microbicides in clinical trial settings, including work with the HIV Prevention Trials Network (HPTN) and CONRAD. She, an active member of the Ethics Workgroup of HPTN, developed an expanded model of the informed consent process that is being used in an upcoming microbicides clinical trial. Dr. Woodsong is currently working with an FHI team to develop approaches to conducting behavioral research in clinical trials settings.

LAWRENCE (LARRY) J. SEVERY obtained his PhD in social psychology at the University of Colorado in 1970. He spent the summer of 1976 at the University of North Carolina participating in one of National Institutes of Child Health and Human Development post-doctoral training programs for population research. His long-term interests have focused upon: couple decision-making regarding family planning and contraception; the acceptability of new technologies and programming aimed at preventing unwanted pregnancies or sexually transmitted diseases—especially HIV; and program evaluation. Dr. Severy is the R. David Thomas Professor of Psychology at the University of Florida, and was the 2002–3 President of Division 34—Population and Environmental Psychology—of the American Psychological Association. Currently, he is the Director of Behavioral and Social Sciences, Family Health International, Research Triangle Park, North Carolina, by virtue of a long-term contract between FHI and the University of Florida.

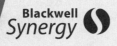

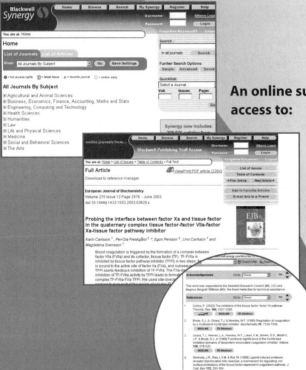